Literary Atlas

This book documents a new approach to literary geographies based around the Literary Atlas of Wales. It introduces an innovative "plotted" approach which empowers reading, creates connections to localities, histories, and communities, and inspires interest in literature and geography.

It showcases how literary geographies can be mobilized through the plotted approach to reading. Through documenting the Literary Atlas of Wales project, this book outlines how the plotted approach was used to engage with English-language novels set in Wales. It argues that the future of this interdisciplinary subject area should be premised upon nurturing instability, turbulence, and experimentation in order to produce new insights which can change the way we understand the relations not only between literature and place but also between other modern categories, including academic disciplines.

This book will be of interest to all readers of literature, human geography, mapping, heritage studies, and tourism. It will be beneficial to those interested in the domains of cartography, creative humanities, cultural sociology, human geography, literary studies, and print cultures.

Jon Anderson is Professor of Human Geography in the School of Geography and Planning, Cardiff University, UK. His research interests focus on the relations between identity, culture, and place, in particular the actions, practices, and politics that such relations produce. His recent publications include a monograph titled *Surfing Spaces* (2023) and the third edition of *Understanding Cultural Geography: Places and Traces* (2021).

Routledge Research in Culture, Space and Identity

Series Editor: Peter Merriman

The *Routledge Research in Culture, Space and Identity Series* offers a forum for original and innovative research within cultural geography and connected fields. Titles within the series are empirically and theoretically informed and explore a range of dynamic and captivating topics. This series provides a forum for cutting edge research and new theoretical perspectives that reflect the wealth of research currently being undertaken. This series is aimed at upper-level undergraduates, research students and academics, appealing to geographers as well as the broader social sciences, arts and humanities.

Literary Atlas
Plotting a New Literary Geography
Jon Anderson

Spaces of Puppets in Popular Culture
Grotesque Geographies of the Borderscape
Janet Banfield

Migration, Community and Identity
Countercultural lifestyle migration to rural Wales, 1965–1980
Flossie Caerwynt

Spatialities of Speculative Fiction
Gwilym Lucas Eades

Becoming Weather
Weather, Embodiment and Affect
Sarah Wright

Coexistence
Spacings, Dis-positions, and Being-with Others
Paul Simpson

For more information about this series, please visit: www.routledge.com/Routledge-Research-in-Culture-Space-and-Identity/book-series/CSI

Literary Atlas

Plotting a New Literary Geography

Jon Anderson

Routledge
Taylor & Francis Group
LONDON AND NEW YORK

First published 2025
by Routledge
4 Park Square, Milton Park, Abingdon, Oxon OX14 4RN

and by Routledge
605 Third Avenue, New York, NY 10158

Routledge is an imprint of the Taylor & Francis Group, an informa business

British Library Cataloguing-in-Publication Data
A catalogue record for this book is available from the British Library

ISBN: 978-1-032-34263-4 (hbk)
ISBN: 978-1-032-34265-8 (pbk)
ISBN: 978-1-003-32124-8 (ebk)

DOI: 10.4324/9781003321248

Typeset in Times New Roman
by Apex CoVantage, LLC

Contents

1 Plotting a new future for literary geographies

Introduction

Over recent years, scholars from a range of modern disciplines have embarked on exciting adventures in the world of literary geographies. This book, *Literary Atlas: Plotting a New Literary Geography*, engages with these adventures, whilst also charting its own journey into these challenging new worlds. The book shares insights from *Literary Atlas*, a project which sought to "plot" new readings of English-language novels set in Wales, and as it does so, critiques the framework in which academic endeavours in this subject area have been understood. More specifically, the book offers a new reading of the contemporary emergence of, and possible futures for, literary geographies. Counter to the growing consensus which suggests that literary geography (notably in the singular) is and should be a discipline forged through the nexus of its two titular subject areas (namely human geography and literary studies), this book offers an alternative reading of the subject. It argues that literary geographies (in the plural) have undergone an explosion of innovation in recent years not as a consequence of the formal coming together of two disciplines but due to the confluence of a range of philosophical changes that have stimulated interest in the subject from a range of subject areas (including, for example, cartography, the creative humanities, cultural sociology, human geography, literary studies, and print cultures, to name several). To this beginning, this book formally counters disciplinary metaphors of "territory" and "turns" to adopt the more "fluid" metaphors of "sea changes" and "currents" to story the context of philosophical change in which literary geographies are mobilizing. The book suggests that, due to the influence of these conceptual changes and currents, literary geographies are far from being a "stable" disciplinary category (after Tally, 2019:292), and are enjoying perfect storm conditions for their continued development. The book argues that the future of this interdiscipline should be premised upon nurturing this instability, turbulence, and experimentation in order to produce new insights which could change the way we understand the relations not only between literature and place but also between other modern categories, including academic disciplines.

The book then presents one innovative output that has emerged from these perfect storm conditions. *A Literary Atlas* harnesses a fluid approach to literary

DOI: 10.4324/9781003321248-1

geographies, combining both "distant" and "scientific" geo-visualizations along with "deeper" mixed cartographies to generate and publish a range of geographical, literary, and multimedia data. As we will see in this book, the *Literary Atlas* project recorded an online "book of maps" which locates over 570 English-language novels set in Wales. Then, focusing on 12 novels in particular, the project created and commissioned a range of innovative cartographies in order to compose a new atlas of literary geographies (www.literaryatlas.wales). To evaluate this range of contrasting and complementary maps, the project embarked on a series of qualitative research activities, involving reading groups, field trips, digital surveys, exhibitions, and public workshops.[1] This book analyzes the insights garnered through both the production of the *Literary Atlas* and the subsequent public engagement with it, to contribute to the ongoing conversation about the potential of literary geographies to change the ways we read, write, and experience literature and place. Through doing so, the book seeks to reflect not only on the significance of contemporary literary geographies but also its possible futures.

Literary geographies: turbulence, diversity, and disciplinary confluence

From the perspective of this book, literary geographies can be generally defined as the subject area which critically explores the complex of literature and its capacities to reflect, restore, and reimagine our relations to the world. Over recent decades, however, literary geographies have undergone exciting and sometimes turbulent times. A growing number of scholars, from a wide range of disciplines, have begun to explore literature as a "primary source material" (Watson, 1983:385). Literature has been activated as a means through which to engage with the human condition and its relations to the places, processes, and people which make up the world. As Brosseau informs us, the subject area now "covers diverse orientations" (2017:11), including critical investigations of the "context[s] in which literary works are produced", the "geographical 'reality' to which they refer", the analysis of the "representations of space" depicted in literature, and how each "literary creation" may affect our understanding of the world around us ((2017). As Alexander (2015) confirms, the diversity of activity occurring under the aegis of literary geographies is such that a range of complementary subject labels are often used to describe its operations, including, for example, "imaginative geography, literary cartography, geocriticism, geopoetics, [and the] geohumanities" (2015:5).

Chapters 2 and 3 will explore this turbulence and diversity in detail, whilst also offering a critical explanation about the consequences they have for how the subject area of literary geographies should be understood and organized. These chapters will identify the emerging consensus that literary geographies is not simply an interdiscipline but a particular type of interdiscipline, one that is "radical" in nature and which is hoped will develop into a singular organization with its own approaches and vocabularies. This book will counter this narrative and suggest that other forms of organizational interdisciplinarity offer more open, diverse, and creative futures for the subject area. Although these alternatives may bring turbulence and experimentation, they nevertheless offer exciting opportunities to re-imagine how disciplinary categories may be operated and imposed, with lessons emergent for a range of subject areas and academic endeavour more broadly. To contextualize this counter-narrative, Chapter 3 will offer a more fluid way to story the development of literary geographies, introducing the metaphors of sea changes and currents – in particular, the currents of culture, spatiality, cartography, and creativity. In the chapters that follow, the *Literary Atlas* project will then be introduced, discussed, and critically evaluated, as one example of a different literary geography which can emerge from this turbulence and experimentation. In order to formally introduce this book's approach to the subject area, and set ourselves up for this journey into a new literary geography, this chapter now continues by directly engaging with the key terms which compose this project's (and the book's) title: "literature", "atlas", and "plot".

Literature, atlas, and plot

In the popular imagination, "literature" or "the literary" often refers to scripts aligned with "high culture". Literature has come to be understood as human representations and practices that have been consistently claimed by influential actors to be the best that has been thought and said, or rather in the case of literature specifically, the best that has been (thought and) written. Literature is predominantly understood as crafted writing, using the repertoire of words and grammar, along with their associated page layouts, orientation, and visual aesthetics, to intentionally represent and communicate stories, places, and meanings. Literature embodies these combined representations to expertly craft communication, demonstrating the capacity of written or printed words to engage an audience. In this way, a characteristic of literature is its ability to attract a diversity and longevity of readership, with each finding reward through immediate or considered reflection with the text. Such reflection promotes a resonant understanding that helps to provide insight into, ultimately, the human condition and our place in the world.

An "atlas", in turn, refers to a collection of maps on a given subject, in the case of this book, for example, English-language novels set in Wales. Definitionally, an atlas refers to a compendium of map materials, which traditionally include illustrations, tables, charts, or textual matter. Maps come in all shapes, sizes, and forms, but are commonly dominated by graphic representation. Atlases, however constituted, often refer to or correspond with a geographic place. All atlases have a common objective, one that complements to some degree the purpose of literature; atlases help orientate the reader to their position in the world and help them (materially or metaphorically) navigate their societal and spatial (i.e. geographical) context.

The book's subtitle, *Plotting A New Literary Geography*, is also instructive. It suggests that the combined aims of the book coalesce to form a new way of *doing* literary geographies. This new way refers not only to the practice of the *Literary Atlas* project itself but also to a new way of considering how contemporary literary geographies have emerged, how they can be conceptualized, and what they can produce. At this stage, it is important to be clear that this subtitle is not intended to suggest that the novelty or "newness" of the *Literary Atlas* project is the *only* way to do literary geographies; indeed, this book argues that the vitality of literary geographies at the present time is the result of a multiplicity of versions and approaches which define its subject area. From the position of this book, there can be no one singular or essentialized *literary geography*, be it in approach, aim, or outcome, there can only be many emergent *literary geographies* (see also, for example, Anderson and Saunders, 2016). However, as literary geographies develop, and their future as a discipline (or even perhaps inter-, post-, or poly-discipline) becomes the subject of debate, this statement gains further significance; it becomes not simply an onto-epistemological position with regard to the nature of literary geographies (and by extension the world through which we are co-constituted) but also a statement about the way in which literary geographies should be (b)ordered and disciplined. It asks how can and should literary geography/literary geographies be conceived and defined, to what end and, to whose benefit? From the perspective of this book, and the *Literary Atlas* project more broadly, a vital way to engage with the possibilities of literary geographies is through the notion of "plot".

The notion of "plot" was introduced in Anderson (2014) as a term which alludes to the entangled connections between words on a page, and our material lives in places. Plot is chosen as a term due to its constituent meanings which serve to identify not simply key aspects that define literary geographies but also the relations between them. Firstly, and perhaps most obviously, plot refers to the beginning, middle, and end (for example) of storytelling. In this sense, plot may refer to dramatic development within a fiction, the arcs of its characters, and the process of conflict and resolution, whilst also referring to genre tropes that authors and readers can acknowledge, anticipate, and enjoy. However, plot does not simply refer to the drama of a story, its second definitive aspect relates to the geographical dimensions in which that story occurs. Drawing on the work of Agnew and Duncan (1989), for example, this geographical dimension may refer to a story's specific location, its cultural locale, or even its sense of place. This geography may also include an emphasis on scale (for example, the intimate and domestic, through to the national or global), along with the cultural associations which combine with these sites. Although in some studies of literature, the geographical aspect of plot may be marginalized or even dismissed as simply the "setting" of a story (a dimension which is "just there" to stage the drama); drawing on authors in cultural geography and (post-)phenomenology, Anderson (2014) insists that these two dimensions of plot are not merely coincidental or passively joined, but rather are directly co-ingredient. For Anderson, the notion of plot insists that we acknowledge the explicit interdependence of drama and geography, recognizing how both are directly affected by each other. The notion of plot therefore demands the recognition of a substantive philosophical approach to the study of fiction (which is returned to in more detail in Chapter 3); namely, the stories within literature cannot be separated from the geographical location in which they are rooted, nor can the geographical locations noted in stories be separated from the literature that includes it. Thus, these two aspects of plot are, as the word itself suggests, directly constitutive of this singular "entity/process" (after Harvey, 1996).

However, this book *Literary Atlas: Plotting a New Literary Geography* develops the notion of plot further. It goes beyond the initial introduction from Anderson (2014) to explore in depth how the geographical dimensions of the term should not be limited solely to the locations which anchor the dramatic dimensions of the story. Location is liberated from the limits of text by embracing a formal change of emphasis in relation to the object of literary geographies: no longer is a "plotted" approach (implicitly or explicitly) focused on literature as the "primary source material" (as Watson noted earlier), it now emphasizes particular *readings* of literature as the key object of inquiry. This book argues therefore that in order to exploit the potential of a "plotted" approach to literary geographies we must demand a focus on "the contingent and generative interaction between book/text, reader and context" (after Thumala Olave, 2021:103). With this emphasis, literature is not judged solely from the perspective of the author (its value is determined by how well, for example, it articulates their intent, or how effectively it purges their need to write in, on, and around that subject). An emphasis on the reading of literature rather reminds us of the obvious but nevertheless significant point that literature, like any other form of culture, realizes a vital part of its potential through engaging with an audience. Literature needs something, or more importantly perhaps in the era of Artificial Intelligence, someone to read it, and it is through this act of reading that the geographical dimension of plot is extended beyond those locations which anchor the fictive dimensions of any story.

An emphasis on the reading of literature explicitly focuses our attention on how an author's plot may conjure new geographies in the mind of the reader. Indeed, it could be argued that literature as a cultural form excels because of its ability to exploit the power of association inherent within human language. Those who engage with literature are drawn into what Watson calls a "special form of reading game" (2000:110, see also Pavlik, 2010), which encourages and often compels them to make associations between the written plot and their own experiences. As Kuzmičová and Cremin put it, this reading game "nurtures intimate connections between [the] text and the [readers' own] life" (2021:37) and as a consequence, read-literature transforms from the author's own words into a story,

> packed with associations from our own personal histories. We see both the thing itself and all that we associate with it.
>
> (Storr, 2019:41[2])

In this way,

> Readers . . . insert their own feelings and interpretations into why that just happened and what it all means. These gaps in explanation are the places in story in which readers insert themselves: their preconceptions; their values; their memories; their connections; their emotions – all become an active part of the story.
>
> (Storr, 2019:55)

Thus a shift of focus from text as authorial artefact to a wider focus on literature as experienced through reading emphasizes how an author's crafted words can excite suggestions in the minds of a reader, and as a consequence, mobilize the generative effect of their "architecture of imagination" (after Larsen, 2018:168). Identifying this process is significant. When an individual chooses to engage with literature, they effectively consent to the possibility that the words chosen by the author will confront, inspire, provoke, challenge, convert, and critique their sense of themselves and the world in some way. The prospective reader is aware that, if the words, grammar, spacings, and paginations are chosen correctly, the act of reading may compulsively engage them in a form of world-building that will combine their own lives with the author's words in unique and unanticipated ways; reading will suspend their disbelief, and they may lose themselves (or should that be find themselves?) in new, combined, geographies, cultures, and action.

This book suggests that shifting our emphasis from literature-in-isolation to read-literature produces a number of key implications for extending the notion of plot, and by extension, literary geographies more generally. Firstly, this shift in emphasis insists that read-literature can no longer be treated as a singular or isolated "thing", any piece of literature must be defined in relation. The experience of reading is fundamental to understanding the nature of what literature is, its meaning conjured and extended, reduced or enchanted[3] through the correspondences that its crafted words provoke in the imagination of its reader. Secondly, a focus on reading insists on the acknowledgement that not all reading is the same. Although it is possible for the act of reading literature to change an individual's relation to the world, it is not inevitable that this will occur. As Thumala Olave notes (earlier), the act and effect of reading can be influenced by context (which may include, for example, a reader's intent, concentration, mood, preoccupation, or place of reading, as well as their socio-demographics in a variety of ways (see, for example, Allington, 2011; Fuller and Rehberg Sedo, 2013)). Indeed, scholars suggest it is possible to explicitly adopt a particular form of reading to engage with a piece of literature. We know, for example, that it is possible to "skim read" a text, or we can scan through for keywords, or even read out loud (paying more attention perhaps to the sound of words, their cadence, and affect). For more scholarly reading, Brosseau (1994) suggests that one can conduct a "close" reading,[4] Moretti suggests a "distant" reading (1998), whilst Leypoldt (2021, after Frank, 1945) suggests it is possible to take a "spatial" reading of a text,[5] with each calling our attention to different generative insights. In line with these possibilities, this book seeks to explore the possibilities offered by grafting the agency of reading to the notion of plot, forming through their conjunction a specific and explicit "plotted reading" of literature. Following Anderson (2014), a plotted reading would be a reading act which prioritizes explicit attention to the co-constitution of drama and place within the text, whilst also seeking to capture and explore the corresponding geographies that are associated with it by the reader. These connections may be to other pieces of literature known to the reader, to other cultural traces (see Anderson, 2021) beyond literature, or to cultural geographies conjured from the crafted word through the reader's imagination. A plotted reading could thus be sensitive to the myriad potential open to the reader to make geographical connections to a story (following Martinez and Herman, 2020), and thus complete the conversion of literary geographies from a narrow study of "an apparently finite textual edifice (the published book in our hands)" to a subject which broadly engages with read-literature as "a potentially infinite hypertext" (Leypoldt, 2021:150).[6] This book explores how this conversion sets in motion new questions for literary geographies with regard to the possibilities of exploring, charting, and mapping the multitude of corresponding geographies enabled by plotted reading.[7]

A plotted reading: before and beyond the modern constitution of the world

As Anderson (2014) noted, within the notion of plot is the possibility for scheming and subversion. When developing the practice of plotted reading, this possibility is made manifest. As we have seen earlier, the notion of plot and the practice of plotted reading explicitly remind us that what we refer to as "literature" is, from this perspective, actually a co-ingredient merger of author's words and reader's worlds. As these two components of literature converge, plotted readings can be seen to subvert the modern tendency to isolate and keep separate these core components of the reading experience. As Latour (1993) has argued (and this book will discuss in detail in Chapter 3), the predominant way in which the world has been categorized in the West is through the "modern constitution". This constitution has, according to Braun, a "proclivity" to "purify the world into essences" (2004:169), cleaving categories from the chaos of the world in order to simplify and enable our engagement with it. One of these key categorizations is the onto-epistemological separation between "texts" (or representations of the world, including literature) on one hand, and "the world" itself on the other. This separation of text and world has presented ongoing problems for literary geographies. Considerable work in this subject area has focused on the complex relations between these two, now separated, ontological domains (e.g. Ryan, 2013, 2015, 2017). Indeed, scholars have had to negotiate this separation even when they may be agnostic with regard to its legitimacy, or even consider it to be a "false opposition" (after Bennett and Royle, 2016:36). As a result, scholars have suggested how literature may "hover above" the real (after Piatti and Hurni, 2011:218), or how it becomes necessary to explore circulations "from the lived to the imagined world, and the imagined to the lived world" (Anderson and Saunders, 2016:117). The normalization of the onto-epistemological distinction between text and world thus presents a foundational problem for scholars working in literary geographies; in Hones' words, it

> presents literary geography with a terminological challenge: even when "imagined" and "lived" geographies are understood to be "seeping" into each other, [because of the modern constitution] there is no easy way to avoid having to name them as separate dimensions, and this tends to undermine the argument for permeability even as it makes it.
>
> (2022a:15)

In the face of this terminological cul-de-sac, this book suggests that the notion of plot subverts (or at least circumvents) the problems posed by the modern constitution. As the notion of plot is defined precisely by the integration of both drama and geography, and the practice of plotted reading by the co-ingredience of crafted word and the corresponding geographies conjured by their reader, it subverts the need to engage with the world as constituted by the modern. From a plotted approach, there is no insistence on separating read-literature from the experienced world, rather it is "taken as read" that words on a page are combined through the process of plotted reading with a range of corresponding geographies which, up until that point, lay beyond them. At the point of its creation, therefore, plotted reading is centred on what this book terms the "reality of the read", a plotted reality which is an ontological plane fusing at the very moment of its emergence what the modern constitution up until this point has separated into the "real" and "imagined".[8]

A plotted reading therefore opens up the possibility to literary geographies that the modern constitution's distinction between "the text" and "the world", and "the real" and "the imagined" can be radically challenged (see Law, 2015). The reality produced through plotted reading does not separate or distinguish these modern onto-epistemological planes, but actively (con)fuses them. It is possible that a plotted reality may only be experienced as a temporary suspension of the modern, enabled by the act of plotted reading[9]; yet through explicitly recognizing its existence, the "associative power" (Storr, 2019:42) of plotted reading is also acknowledged and, as Bennett and Royle insist, we realize afresh how

> texts cannot but be part of the world. To talk about texts as "representing" reality simply overlooks ways in which texts are already part of that reality, [as well as the] ways in which literary texts *produce* our reality.
>
> (2016:36)

Into the borderlands

With this capacity of plotted reading in mind, we return to the full title of the *Literary Atlas* project: *A Literary Atlas of Wales and its Borderlands*. As we approach the final stages of this opening chapter, and in light of the promise that plot and plotted reading offers to literary geographies, the reference to "borderlands" becomes centrally significant. At face value, its inclusion suggests the wish to ensure that any approach to the geographical scale of the project – the nation of Wales – is not reduced to an insular, bounded, or sedentary framing (see Chapter 3) but rather is open to a processual, connected, and outward-looking state of the nation. However, in light of the significance of the project's enactment of plot (in all its dimensions), the reference to borderlands has further consequence. Following the author Phillip Pullman's use of the term, borderlands can be seen to describe the plotted readings which are prompted through the reader engaging with the words of a fiction, and associating them with their own range of corresponding geographies. For Pullman, these merged geographies are "a borderland"; in his words, this borderland is the literary geography,

> that opens up between the private mind of the reader and the book they're reading. It'll be different for every individual, because while parts of the borderland belong to the book, other parts belong only to that particular reader – to us: our memories, the associations we have with this or that particular word or landscape, the aspects that resonate with our own individual temperament; so whereas many readers might be reading the same book, no two of them will read it in exactly the same way. However, we can come back and talk about it; we can tell others about our experiences of it, and compare our part of the borderland with other people's.
>
> (2017:259)

This book outlines how, in the spirit of reading, sharing, and exploring that Pullman suggests, the *Literary Atlas* project sought to critically explore a new way to read, map, and produce literary geographies. It suggests how the *Literary Atlas* project sought to not only "plot" the reading of English-language novels set in Wales but also attempt to engage with the borderlands that were inevitably conjured up when a "plotted reading" was encouraged. Committing to explore the worlds created through a plotted reading, and seeking to create a range of maps to chart the corresponding geographies convergent through this process, produced an approach which offered a new vantage point from which to understand the practice of literary geographies. This book suggests that the *Literary Atlas* offers the opportunity to suspend the modern limits of the subject area and open up new possibilities for plotting a new world of literary geographies.

Book outline

As stated earlier, Chapters 2 and 3 will continue by exploring the current state of literary geographies. Chapter 2 will outline the growing consensus which suggests that literary geography should develop as a particular form of interdiscipline, one that is actively seeking definition around a relatively narrow set of implicitly modern conventions. In contrast, Chapter 3 offers a counter-narrative for literary geographies, emergent from a new way of conceptualizing the relational philosophies that have influenced the social sciences, arts, and humanities over recent decades. Chapter 3 makes the case for the significance of a "sea change" in amodern philosophical approaches which, it argues, have set in motion an array of conceptual currents (in culture, space, cartography, and creativity). It will argue that together these have produced perfect storm conditions for literary geographies, and the chapter concludes by offering its own definition for a diverse, plural, and inevitably turbulent iteration of literary geographies as a "functional" interdiscipline.

From this basis, the remainder of the book will present and then evaluate one iteration of literary geographies emergent from this definition. Chapter 4 will begin this process by introducing the motivations behind and the plans for the *Literary Atlas of Wales and its Borderlands*. Chapter 4 will outline how the *Literary Atlas* drew on a range of bibliographical and publishing databases to collate an overarching cartographic catalogue of English-language novels set in Wales, and how, based on this catalogue's diversity in genre, geography, history, and authorship, selected 12 novels to form the basis of more comprehensive mapping activities. Chapter 5 outlines how *Literary Atlas* went about a "plotted reading" of the texts, moving from distant maps to deep maps of the borderlands of each novel's literary geographies. For each novel, detailed localities information was produced (including Census data, local histories, and cultural events, in statistical, visual, and written forms) and was further contextualized through the addition of author and/or expert commentaries (from first-person interviews), narrative excerpts, historical and contemporary photographs, as well as film and sound recordings. This re-contextualization transformed "distant geo-locations" into in-depth "plot-points", entangling the literary plot with its historical, geographical, and cultural assemblages. Chapter 6 goes on to outline how direction and meaning were given to these cartographies through composing a range of "plotlines" (see also Anderson, 2016a). Plotlines were new routes through the plot-points of these newly mapped literary geographies, and offered the potential to open up new excursions into these borderlands (see Chapter 11). Chapter 7 then outlines how this combination of distant maps, deep maps, and plotlines was used to inspire a range of cartographic artworks. Professional artists were commissioned to interpret the maps generated by *Literary Atlas* from a plotted perspective, and re-articulate the relations between culture, place, and literature that they suggested. These artworks, combining canvas, sculpture, music, and film, were curated by the *Literary Atlas* team and exhibited at 6 venues across Wales.

As Dewsbury notes, any relational assemblage "is less about what it is . . ., and more about what it can do, what it can affect and bring about" (2011:150). In this spirit, the remaining chapters of the book directly outline and analyze the responses gained through a multiplicity of public engagement activities with the *Literary Atlas* materials. Firstly, Chapter 8 engages with authors and experts to discuss how a plotted approach to literature can be used to illuminate the writing approach to the novels centred on in the *Literary Atlas* project. Then, the book turns its focus to an evaluation of the practice of plotted reading. 21 reading groups across Wales were recruited to examine how the *Literary Atlas* changed the ways in which public readers engaged with fictions and their geographies. Chapter 9 explores the first stage of this engagement, outlining reading groups' responses to the plotted reading of their selected novels. Chapter 10 details the effects of the plotlines created by *Literary Atlas*, and how they influenced the reading and peer groups' understanding of the novels, the locations, and their co-ingredience. Chapter 11 discusses the third stage of reading group engagement, outlining the effect of taking individuals into the "field" and mobilizing the virtual "plotlines" in practice. These plotted excursions involved the reading groups engaging with and further contributing to their real-realities through direct engagement with borderland spaces. Chapter 12 details and analyzes the affects set in motion by the cartographic artworks and exhibitions.

Over 20,000 individuals attended these exhibitions, which occurred alongside public workshops which targeted "hard-to-reach" members of the public. Chapter 12 evaluates the merits of these innovative cartographies in promoting understanding and awareness of literary geographies. Chapter 13 brings together these evaluations to address the inter- and cross-disciplinary insights from the *Literary Atlas* project. It discusses the direct conceptual, methodological, and practical benefits for the range of disciplines which converge to form literary geographies, offering insights into how the re-invention of literary mapping can offer the potential for new synergies between mapping activities and amodern approaches. The chapter demonstrates how the *Literary Atlas* project moves the category of literature forward into an intellectual, imaginative, and embodied encounter, and develops literary geography away from "simply" being the cartographic (re)production of fictional texts, into an interdisciplinary mode of operation which examines the borderlands of read-realities. As it does so, the chapter concludes by suggesting how academia can fulfil a public role by mobilizing interest in literature, history, and culture, and stimulate public participation in these activities. It suggests that literary geographies should grasp the moment for developing new approaches, like plotted reading, which can mobilize literary and cultural resources to be of social and economic benefit to local, regional, and national communities.

Notes

1 The full title of the project was *A Literary Atlas of Wales and its Borderlands*, and was sponsored by the *Arts and Humanities Research Council* (UK). The author was the Principal Investigator on the project, and led a project team including Professor Kirsti Bohata, Dr Kieron Smith, and Dr Jeffrey Morgan. Professor Scott Orford and Professor Damian Walford Davies acted as Mentors to the project. Literature Wales, the national company charged with developing literature in Wales, was a project partner. The *Literary Atlas* website (www.literaryatlas.wales) was generated by the author and the key project team; this book and its arguments were developed, curated, and written by the author.

2 Storr's full quotation here continues as follows:

> We feel it too. Everything our attention rests upon triggers a sensation, most of which are minutely subtle and experienced beneath the level of conscious awareness. These feelings flicker and die so rapidly that they precede conscious thought, and thereby influence it. All these feelings reduce to just two impulses: advance and withdraw. As you scan any scene, then, you're in a storm of feeling; positive and negative sensations from the objects you see fall over you like fine drops of rain. This understanding is the beginning of creating a compelling and original character on the page. A character in fiction, like a character in life, inhabits their own unique hallucinated world in which everything they see and touch comes with its own unique personal meaning. (2019:41)

3 It could be argued, for example, that literature excels as a cultural trace due to its ability to exploit the paradox of representation (see Bondi, 2005a) in order to "enchant" the reader (see Woodyer and Geoghegan, 2012), a point we will return to later in this chapter).

4 A close reading, in the spirit of discourse analysis, contemplates the precise contents of literature and these can generate insights into its dynamics and political intent (see also Chapter 3).

5 Spatial reading as it is conceptualised here follows in the tracks of Frank (1945), identifying the method as a subset of scholarly reading that does not "begin and end in the pleasure of consumption" but must give rise to "sustained reflection" (Guillory, 2000:31).
6 Connecting this idea to the notion of plot explicitly, the term's Latin root *cosentio* refers to the practice of agreeing, consenting, and even conspiring, to act together. In this case, therefore, plot can also be understood to refer to the reader agreeing to act with the author's words, and conspiring with them to form the story.
7 We will examine this point further in Chapter 2 through the work of Hones (2014), Thurgill and Lovell (2019), and Thurgill (2021).
8 The prevalence and power of what this book terms the "reality of the read" is a phenomenon that has been widely noted by writers of fiction. As Jess Burton puts it, "stories [can often feel] more real [to me] than life" (Burton, 2023:no page), whilst Alan Garner insists that read-literature can feel "more real than the furniture of this room" (2010:138).
9 When reflecting on his own reading practices, Phillip Pullman states that although he "knew [with his modern mind that fictions] were made up", "they could [still] be . . . real as well [. . .] in order [for me] to enjoy being with them" (Pullman, 2017:311). Academic and writer Andy Martin has also witnessed this phenomenon, as he notes when commenting on how the author Lee Child composes fiction: "When he [Child] writes, he goes into a 'zone' in which he really believes that the non-existent Jack Reacher is temporarily existent. 'I know I am making it up' [Lee says], 'but it doesn't feel that way'" (Martin, 2020:19).

2 Literary geographies and interdisciplinarity

As we have seen in Chapter 1, this book argues that scholars from a range of modern disciplines have come together to redefine and recompose the world of literary geographies. Yet, as interest has grown, competing claims over what the nature of this subject area should be have also arisen. Indeed, the passion and commitment individuals have shown to the practices, aims, and possibilities of literary geographies have sparked a battle over who can "discipline" the subject area, both organizing and controlling what can and should "be accepted as good practice" (Hones, 2022b:18). This chapter uses insights from Donaldson et al. (2010) to identify these competing claims, and the consensus that appears to be emerging from them. It will show that, from an initial acceptance of diversity and plurality, literary geographies have been reduced to a particular form of literary geography (in the singular), occurring in the space between its purported key disciplines: human geography and literary studies. Although such disciplining may help to counter precarity in the short term and accelerate academic empire-building over a longer time frame, it also forgoes the opportunity to explore the creative tensions produced by operating in a "multi" and "functional" interdisciplinary context. The chapter will conclude by briefly showcasing the vitality of retaining a functional interdisciplinarity with reference to two terms that have been used to set the foundations for the new discipline: "event" and "hinge". It does so not to critique (or undermine) these terms per se, but rather to argue that their potential can be extended by positioning them within a "functional" interdisciplinarity rather than a narrow version of literary geography. It suggests therefore, that it may be possible, perhaps even desirable, for this particular version of literary geography to coexist with other more "functionally" defined literary geographies in order to retain the overall vitality of the interdiscipline(s).

Interdisciplines

In the twenty-first century, scholars from a range of disciplines have been mobilized by the possibilities offered by engaging with the subject of literary geographies. Initially, this coming together of approaches was welcomed by those involved,

DOI: 10.4324/9781003321248-2

with interests expressed in exploring this plurality of insights and their potential interaction. As Jones argued,

> It is my contention that we should embrace the broadest possible scope [of literary geographies], accepting the inherent, perpetually enriching interdependence of space and literature.
>
> (Jones, 2018:40–41)

The sentiments of inclusivity and interdependence were reflected in the opening issue of the new journal for the subject, *Literary Geographies*. As Alexander proclaimed,

> The plural form of this journal's title, *Literary Geographies*, is intended to accommodate and encourage such diversity.
>
> (2015:5)

The term "interdisciplinarity" came to be adopted as a term to identify and story this common space, being used by a range of scholars who have come to define the subject area (see, for example, Alexander (2015), McLaughlin (2018), Hones (2022a), as well as my own work, e.g. Anderson, 2014). As Alexander identifies, literary geographies came to be accepted as "an emergent *interdisciplinary* field of research" (2015:3, emphasis added). Explicitly adopting the notion of an interdiscipline was a useful manoeuvre; it formally recognized that literary geographies did not fit into existing disciplinary categories. As Chapter 3 explores in detail, work in this area was more than the subject of literature as an isolated text, more than the study of how literature interacted with space however conceived, it did not fit neatly into the category of cartography, or was not easily reduced into simply artistic creativity. It therefore appeared to be a subject area that modern faculty-based categories could not contain (existing beyond the limits of science, social science, arts, and humanities). It was accepted that literary geographies could not be appropriately defined with reference to any modern disciplinary category.

For scholars working in literary geographies, identifying this lack of academic homeplace (after Hooks, 1991) was important: it affirmed their awareness that their work rarely fitted within established modern conventions, with concomitant problems of recognition, resources, and precarity. The use of the term interdisciplinarity was thus a useful political manoeuvre; it recognized and responded to the betwixt and between nature of literary geographies, whilst also promising an exciting future. It offered the potential of a yet-to-be-defined common space of operation where disparate scholars could convene to acknowledge diverse origins, recognize creative differences, and forge new alliances.

Despite the widespread use of interdisciplinarity to describe the nature of literary geographies, it nevertheless remains a practice that means different things to different people. Interdisciplinarity can refer to a variety of scholar endeavours, including working practices, institutional set-ups, aims, methods, philosophical influences, and outputs (see, for example, Barry et al., 2008 and Szerszynski and Galarraga, 2013). As Pasquini et al. note, the complexity inherent within the term and the lack of common understanding about its depth and nature is significant, as they state:

> The profound degree to which researchers' assumptions, expectations and attitudes (which in turn are affected by their personal backgrounds, training, location, etc.) influences the very notion of interdisciplinarity, and what it involves and consists of, is often neither noted nor appreciated.
>
> (2008:552)

In order to understand this complexity, Donaldson et al. identify "different degrees of depth to interdisciplinary working" (2010:1523). They offer a threefold typology to distinguish these differences. The first type of interdisciplinary they identify is effectively a form of multidisciplinarity and involves scholars (perhaps briefly) emerging from their own silos of interest to operate "in discrete disciplinary work packages that are subsequently stitched together with little meaningful cross-disciplinary engagement" (Donaldson et al., 2010:1523). The second type of interdisciplinarity is deeper and more "functional" in its nature (after Donaldson et al., 2010). It involves a commitment to "data exchanges" and the development of "common epistemological approaches" which explicitly seek to function together to "link different disciplines", and work towards "framing *integrated* research projects" (Donaldson et al., 2010:1523, my emphasis). It is important to note that, despite functional interdisciplinarity engaging with and reciprocally informing the different conceptual approaches that have come together to form the project in question, this depth of interdisciplinarity does not guarantee each discipline equality in their relations. It is often the case that one discipline retains dominance in these relations, with other(s) becoming subservient (after Barry et al., 2008). The third and deepest form of interdisciplinarity according to Donaldson et al. (2010) is more "radical" in nature. This interdisciplinarity (what is sometimes termed transdisciplinarity (Barry et al., 2008) or "bridge-building" interdisciplinarity (Klein, 2010)) has the aim of bringing together two existing modes of operation to form a new disciplinary "paradigm" (after Szerszynski and Galarraga (2013:2819, invoking Kuhn, 1970); in short, it seeks to create a new disciplinary space. It is tempting to suggest that these three types of interdisciplinarity form a linear process, with increasing depth extending over time; indeed, in the case of literary geographies, this chapter will argue that in the contemporary era, the subject area was initially defined by multidisciplinary working, then evolved into different variations of functional interdisciplinarity, and more recently its "main practitioners" (after Soitu, 2021:115) have established a bridge-building interdisciplinarity which is emerging as the new singular "paradigm" for literary geography. However, as stated earlier, this chapter also suggests that this process should not be seen as linear, inevitable, or even useful, as the potential offered by a looser, functional interdisciplinarity may offer the greatest potential for creativity and innovation in the future.

Literary geographies as multidisciplinarity

In the contemporary period, it is possible to identify that much work in literary geographies is "multidisciplinary" in form. As noted earlier, multidisciplinarity involves scholars contributing their respective ontological, epistemological, conceptual, and methodological expertise to answer particular questions in a broader project, but perhaps falling short of synthesizing their results or looking for insights across the range of questions the project engages with. On the face of it, this siloed approach may not be a problem; if projects do not require formal synthesis, or particular work packages can thrive in semi-isolation, multidisciplinarity can allow interesting and exciting work to flourish. However, when one considers the need to review or assess any project in aggregate, such multidisciplinarity renders more complex the problem of identifying appropriate expertise. It is not always the case that every reviewer, for example, can claim working knowledge in a multiplicity of disciplinary approaches, or that experts in one area will accept the disciplinary assumptions of a different subject area. These problems have been identified by Hones in her role as a lead editor of the journal *Literary Geographies*,

> the journal *Literary Geographies* . . . [has faced problems in] both the peer-review and the copy-editing stage . . . in interdisciplinary communication, particularly when an author is using terms and concepts with obvious connotations in their own field or native language but which do not function in the same way for readers trained in a different field.
>
> (2022b:19–20)

Hones goes on to explain how this problem might manifest itself in practice:

> This means, to take a practical example, that apparently simple and self-evident terms such as space or reader have to be carefully used. . . . A literary critic might well use the term place as if its meaning is self-evident; a cultural geographer probably would not. Meanwhile, a cultural geographer might well refer to the reader as if that concept, too, were unproblematic; a literary scholar working in reception studies probably would not.
>
> (Hones, 2014:8)

Although it is important not to "underestimate" (after Hones, 2022a:18) the differences at play here, it is also important not to overestimate them. In the contemporary context, the disciplines from which literary geographers emanate do not exist in isolation, and those operating within them have not been immune to the conceptual currents of culture and space (see Chapter 3) that have washed across the social sciences, arts, and humanities for three decades. As a consequence, it is unlikely that the majority of scholars working on the subject of literary geographies, whatever their disciplinary homeplace, will be unaware of the theoretical complexities of place or uncritically refer to a disembodied "reader" (for example). However, whatever the magnitude of these problems of aligning expertise may be, they can nevertheless be addressed in a number of ways. Those editors and reviewers who do exhibit multidisciplinary knowledge can be identified in order to appropriately engage with project or journal submissions, whilst multidisciplinary projects can do more to synthesize their approaches in line with funding body criteria or journal guidelines. If it is deemed desirable, it is likely that with advice, example, and encouragement it would be increasingly possible to formally speak across disciplines, a process that may help to produce a different type of interdisciplinarity over time.

Functional interdisciplinarity in literary geographies

Indeed, a number of scholars working in literary geographies have sought to encourage a deeper, more functional, interdisciplinarity in this subject area. Robert Tally (2017), for example, coined the label "spatial literary studies" to describe his version of this activity, emanating directly from the discipline of literary studies. In a later paper, Tally described spatial literary studies as

> a way of characterizing a wide variety of work in literary studies that paid attention to space, place, mapping, spatial relations, geography, architecture, and so on.
>
> (2019:392)

With Donaldson et al.'s typology in mind, spatial literary studies could be seen as a form of functional interdisciplinarity, where more than one discipline (here literary studies) seeks to integrate with others (namely, geography, and perhaps cartography and architecture) to actively reflect on the insights gained from these integrated practices. Functional interdisciplinarity, as a deeper form of combined working when compared to multidisciplinarity, seeks to critically reflect on the tensions and complementarities arising from combining disciplines, and open conversations with scholars from each about mutually beneficial ways forward. However, as Donaldson et al. note (earlier), such functional interdisciplinarity runs the risk of unequal power relations between the combining disciplines. In the case of spatial literary studies, Tally's initial interest emerged from literary studies, with the impression given that the other disciplines were brought in to supplement and serve this dominant perspective. The potential inequality of this partnership was acknowledged by others working in literary geographies, including Sheila Hones, as she identifies:

> Robert T. Tally Jr.'s editorial introduction to the *Routledge Handbook of Literature and Space* (2017), . . . takes literary geography out of its human geography context to incorporate it into the "growing body of work in spatial literary studies" (Tally, 2017:2).
>
> (2018:146)

The point that Tally's iteration of literary geographies ran the risk of prioritizing literary studies over other disciplines may not necessarily have been a problem in and of itself. Given the lack of formal disciplinary setting for literary geographies, and the acknowledgement that scholars approached this space from an array of subject areas, it may be inevitable that scholars exhibit at least a passing allegiance to the discipline (and associated publishing cultures) that employ them. Nevertheless, it is possible to debate the nature of the relations between the respective disciplines that come together in any version of functional interdisciplinarity. In the spirit of inclusion and diversity noted earlier, it can be argued, for example, that literary geographies should be a space in which a range of disciplines can come together and meet as equals, rather than the first, second, or third discipline (etc.) claiming proprietorial rights over others. Perhaps surprisingly, however, Hones did not take this position in response to Tally's version of interdisciplinarity. In contrast, Hones argued that it was not for literary studies to claim literary geographies as "their" interdiscipline, but rather it was for human geographers to claim the subject as theirs (2022a:44). For Hones, literary geography should be an interdiscipline whose dominant disciplinary partner was human geography (and if there were to be subordinate relations, literary studies should fulfil this role). As she states, it should be accepted that

> literary geography . . . has developed from its human geography origins in collaboration with work in literary studies.
>
> (2022a:44)

It became apparent, therefore, that Hones had framed Tally's move to create his own form of functional interdisciplinarity as an act of "annexation" (2018:147). In doing so, the impression was given that the diversity and openness of literary geographies were now subject to a territorial dispute, with actors seeking to establish their own "patches of intellectual ground" and actively "defending" these from what they saw as "illegal immigrants" (after Becher, 1989:24, see also Sayer, 2003, in relation to wider disciplinary disputes across the social sciences). Despite being apparently open to the possibility of (some form of) interdisciplinarity, this emerging "turf war" (after Soitu, 2021) suggested that these "main practitioners" (Sayer, 2003:115) had not "actually escape[d] from their home disciplines" (Sayer, 2003:5[1]), or more broadly, the associated modern tendency to align academic progress with colonization and empire-building. Although such skirmishes offered the possibility for *either* literary studies *or* human geography to gain new "ground" for their interest, it also had a negative effect. Now the context for debate was set: any scholar seeking further engagement with literary geographies now had to demonstrate allegiance (either explicitly or implicitly) to one or other version of this functional interdisciplinarity (for example, in terms of their journal- or citation-choices). As Tally lamented,

> I would hope that researchers in various fields who ought to be allies would not go out of their way to start such wars, even as the impulse to define one's own territorial boundaries may be powerful.
>
> (2019:405)

Radical interdisciplinarity: an emerging consensus on a singular literary geography

Perhaps in response to this turf war, Hones (2014, 2022a, b) offered what can be seen to be her vision for a more fixed and "radical" future for literary geography as an interdiscipline (after Donaldson et al., 2010). As we will see, Hones' radical interpretation of interdisciplinarity was a "bridge-building" exercise between what, in her view, were the two key disciplines which defined the subject area: human geography and literary studies. Hones argued that, from the space between these two disciplines, a new paradigm of operation could emerge. As she states,

> Literary geography, as an academic crossover field, is something else that has to happen in "the space between", in this case, in the gap dividing and connecting literary studies and academic geography.
>
> (2014:4)

Through the use of the phrase "the space between", Hones explicitly recognized the power of creativity "in the space between" a number of key actants, including the reader and the author,[2] but also in this case between literary studies and academic geography (2014:4).[3] Although she did not explicitly draw on the work of Soja here,[4] her recognition of the potential for innovation beyond and between two heretofore apparently stand-alone entities echoes Soja's notion of thirdspace (1996). For Soja, thirdspace refers to a new location which actively seeks to

> break down the categorically closed logic of "either/or" in favour of a different, more flexible and expansive logic of the "both-and-also".
>
> (Soja, 1999:258)

As such, Hones' "space between" can be seen to serve a similar purpose to Soja's thirdspace, operating as a conceptual device, a political concept, and a narrative script; on all these levels, it becomes a beguiling response to any (categorical) "confinement" (to use Soja's terms, 1999:267). Hones' creative space between these two established disciplines (which could be understood as an interdisciplinary thirdspace) opens up a new frontier for literary geographies, offering exciting new alternatives, and the possibility for a creative conjuring "that draws selectively and strategically from the two opposing categories to open up new alternatives" (Soja, 1996:5).

The parallels with Soja's notion of thirdspace become more obvious as Hones adopted a metaphor to extend her framing of literary geography. This was not to be simply the creative production forged between two disciplines, but between two "towers". In her words,

> [literary geographies] is a work that is performed, as we might say, in the space between the tower of geography, on the one side, and the tower of literary studies, on the other.
>
> (2014:4)

In Hones' radical version of literary geography, the subject should be a conjunction of two disciplines, literary studies and geography, and these disciplines can be understood as "towers". It is important to emphasize this point as such allusion suggests that each discipline is "modern" in its nature (after Latour, 1993, see Chapter 3): coherent, singular, and united in its ambition, and wholly separate from the other. In Hones' words, these contrasting disciplines amount to "two well-established structures, with independent foundations, which afford different views" (2014:4). From this perspective, therefore, disciplines are architecturally solid and perhaps monolithic in their character – suggesting (although not explicitly arguing) that those inhabiting them share the same sense of home; these are disciplines which house a "community of scholars with a shared vision of their science's goals and methods" (Johnston, 2000:974) (rather than the counter-view which sees disciplines as housing "quasi-independent communities", with scholars existing within them who experience "a substantial degree of isolation from others within the[ir] discipline" and enjoy "stronger links to others [outside it]" (Johnston, Johnston, 2000.)).

To some extent, therefore, Hones' independent towers of literary geography conform to the modern notion of a discipline, operating monolithically, autonomously, and more or less homogeneously. Whilst doing so, Hones' towers also "build" on commonly accepted modern clichés of academia as operating in "ivory towers", positioning academics above and outside their study, perhaps in accordance with an Archimedean positioning (and thus the metaphorical language suggesting a philosophical stance and material location). More particularly though, as we know, for Hones, her version of literary geography does not occur *in* these towers, but in the space *between* them. This operation is concisely explained by Tally when he states,

> Hones use[s] the evocative metaphor of the tightrope walker . . . in order to emphasize the interdisciplinary nature of [her] literary geography . . . The literary geographer, in this view, may engage in work that is associated with two disciplines, but this work is also lies outside of both . . . In this vision, the perils as well as the rewards of interdisciplinary research are highlighted, as literary geography is defined as a balancing act that requires equal participation in distinct disciplines of which one cannot wholly be a part.
>
> (2019:402)

This view reiterates the argument that literary geography can only exist in the creative space between literary studies and geography; yet now this creative space is not down in the trenches (if we adopt a terrestrial metaphor), or in the confluence of currents (if we turn "blue" in our language, see Chapter 3), but remains above and abstract: literary geography is defined and confined by the high-wire strung between these towers. As such, this metaphor at once romanticizes the literary geographer as adventurer, acrobat, or stunt operator – yet whilst it does so, also delimits their activity. In this framing, literary geographers can only work in the strict area defined by these two disciplines, and on a presumably horizontal plane linking particular storeys of these buildings (although it may be possible to establish many wires between many different storeys). In a sense, therefore, these two disciplinary towers – each with their "different views" – become towers from which the subject is surveilled; towers become watchtowers, wires become a space of patrol and for control, in effect, the subject area is disciplined under their gaze.[5] Very quickly, the metaphor emphasizes the key role of (just) two disciplines, and runs the risk of converting the openness and flexibility of the creative space of literary geography into a territory to be colonized, enclosed, then defended.

This risk of enclosure is emphasized further if we turn to the specifics of how Hones suggests literary geographers on the tightrope should operate. According to Hones' prescription:

> *Literature* and *geography have* [my emphasis] to function as a combined double subject [with a] combined, double, theoretical, and methodological framework.
>
> (2014:8)

For Hones, this "double subject" has explicit definition; the "defining characteristic" (2018:146) of her version of the subject:

> is its double interdisciplinarity: the "literary" of literary geography refers both to literary texts and to literary studies, while the "geography" of literary geography refers not only to real and imagined geographies but also to human geography as an academic discipline.
>
> (2018:146)

For Hones, literary geography is defined by this "doubling down" on the only two subject areas that are relevant to this work: literary studies and (human) geography. "Good" literary geography in this view might only be put into practice by a scholar equally committed to "nurture an explicit collaboration between geographers and literary studies scholars" (2022b:22) or by "strictly collaborative efforts by teams comprising literary critics and geographers" (Tally, 2019:403).

Thus, in a small number of definitional manoeuvres, literary geography (now in the singular) has been shaped from a broad multi- or functional interdisciplinary space into "one" radical new discipline, strictly defined by space between just two parent disciplines, and limited to particular relations between them. To consolidate this position and recruit a consensus, calls to action (or should that be arms?) were made. Hones called for scholars supporting this version of literary geography to generate a more widespread historical awareness of this double disciplinarity (see Hones, 2014), and second, impose new lexicons to describe and story its future developments (see Thurgill, 2021). However, almost as soon as consensus began to emerge in support of this "radical" iteration of (inter)disciplinarity, notes of caution were sounded. Doel, for example, invoked the ideas of Derrida (1989:73) (a philosopher perhaps pointedly operating beyond the confines of human geography or literary studies) to suggest that any process which seeks to determine "what is" (and what is not) literary geography would inevitably be a messy process, exclaiming:

> Surely you didn't imagine that [cutting up and disciplining] literary space would smell like a rose garden: an abattoir more like! Meatpacking and "sense-packing" have much in common, from the disassembly line to the rendering plant. Space – splayed out, disassembled, and rendered.
>
> (Doel, 2018:45)

Similarly, Soitu suggested that any disciplinary "splaying" would need to be undertaken with the full awareness that it used a double-edged sword:

> Marking borders could prove to be either useful, as a means of theoretical clarification, or alienating, therefore creating adversity and territorial seclusion.
>
> (2021:116)[6]

This book suggests that introducing a radical version of interdisciplinarity for literary geographies, what in effect is a new singular discipline (perhaps a Honesian Literary Geography?), is an alienating and unnecessary move. Not all scholars working in literary geographies (in the most plural and diverse sense) emanate from, or feel appropriately housed in, Hones' double disciplines (human geography or literary studies), nor is the subject solely influenced by ideas emanating from just these two areas.[7] The choice to conceive disciplines as "towers" in this context becomes a provocative and limiting metaphor, whilst seeking to fix a new discipline within a fragmenting and perhaps even eroding disciplinary "landscape" (see Chapter 3) appears to show a lack of awareness of moves away from modern categories and a groundswell of interest towards post-disciplinary alternatives (see, for example, Pernecky, 2020).

Beyond these broad concerns, this chapter continues by examining two of the key concepts which have been introduced to provide a new vocabulary for a double-disciplined literary geography: "the hinge" and "the event". This is not undertaken to critique these terms per se, but rather to demonstrate that their potential is far greater if positioned beyond the limits of a narrow literary geography. By bringing these terms into conversation with others that emanate from disciplines beyond literary studies and human geography, but nevertheless share their linguistic origins, the chapter will show that diverse and plural literary geographies offer potential to develop new concepts that could be missed within any narrow iteration of the (inter)discipline.

In a context of functional interdisciplinarity: the potential of "the event" and "the hinge"

Amongst a range of scholars who began promoting an emerging consensus on a Honesian literary geography (see, for example, Hones, 2014; McLaughlin, 2018; Thurgill, 2021), James Thurgill writes that

> As an interdisciplinary project primarily led by theory and practice in human geography and literary studies, literary geographers have borrowed much of

> their terminology from these contributing disciplines. But, as this field continues to develop, scholars working in the area require their own vocabulary.
> (2021:152)

This chapter suggests that two key terms that have emerged as part of the vocabulary for this new version of the discipline are "the spatial hinge" and the "spatial event". As will be noted in Chapter 3, with her notion of the "spatial event", Sheila Hones emphasizes the intrinsically spatial nature of any piece of literature (2014). According to Hones, a book "happens" through an "intermingled processes of writing, publishing, and reading" (2014:19), and at once, "a reader tak[es] up a text, [they] have set off an event" (2018:1301). As she explains, when a reader reads,

> a moment of text-based spatial interaction [is brought into being; the writer, their words, and the reader's interpretation of them are] engaged across distance, participating in an improvisation that is bringing together a broad array of people, places, times, contexts, networks, and communities.
> (2008:1301)

Reflecting the cultural turn (or current as it is defined and explained in Chapter 3), this framing suggests that fiction is not completed when it is written, but this apparent end point simply begins a new process which involves editing, translating, proofing, type-facing, designing, marketing, and positioning in a (virtual or material) shop or library, before the public at large engages with the fiction through the act of reading and its subsequent generative practices. At each of these stages, the literature "happens"; in other words, the literature comes into being: it "occurs". However, beyond this basic sense of being "eventful", Hones' concept refers to the text also happening in a more extended, complex way; to use the terminology of cultural geography (see, for example, Anderson, 2010, 2015, 2021), the moment a reader takes up a text, this interaction "takes and makes place". The interaction brings about not simply a social happening, but also a spatial one, connecting a "broad array of people, places . . . [and] contexts . . ." (see Hones, earlier) in the broadest sense, and, through so doing, invents and influences a range of (what are now literary) geographies.

The notion of literature as a spatial event thus conceptualizes the interaction between writer, book, and reader (for example) over times and spaces, creating an idiosyncratic network of connections over multiple dimensions (in the style, perhaps, of a "4-D" detective's "crime board", see Johnson, 2017[8]). The notion of literature as a spatial event thus offers great scope to scholars working in literary geographies, but it is also important to note that the notion of the spatial event (although a precursor to Hones' own iteration of literary geography) does prefigure her particular vision for the (inter)discipline. Indeed, the spatial event can be read as a key theory created on and for the "highwire" which connects the two towers of a double-disciplined literary geography.

We can see the influence of the two towers of literary geography through Hones' suggestion that the most appropriate means through which to identify the effects of a spatial event is through reference to the "text". In more detail, she suggests that firstly, spatial effects may happen in the "intra-textual" geographies of the story, in other words, within the text itself (2014:101). Secondly, spatial effects may occur in "inter-textual" literary spaces, that is, in other published fictions beyond the immediate story. Hones also identifies that invention and influence can occur beyond the "text", in other words, to "extra-textual" or "real world" spaces (through, for example, practices of tourism or marketing, or by playing "a major role in how that place is experienced" (2014:67) through the individual experience of a specific reader).

Choosing to define the effects of a spatial event in terms of the "text" may seem an unproblematic choice – a simple adoption of an accepted and normalized shorthand for distinguishing between the "make believe", "fictional", and the "imagined" on one hand, and "truth", "fact", and the "real" on the other (see Chapter 1). However, its employment can be argued to have a number of consequences. By emphasizing text as the key reference point for understanding literature's influence, Hones could be seen to be emphasizing the importance of literary studies (and its disciplinary predisposition towards textuality) on one hand; then, by referring to the "extra-textual", Hones is calling specific attention to the role that human geography and its traditional foci of interest has to offer (the world beyond the text) on the other. The implication is clear (and in keeping with Hones' double disciplinarity): these two towers are the supposed key positions from which to chart the influence of literature as a spatial event; literary studies dominates the territory of the text, human geography champions the word of the world, and it is only from a double-disciplined vantage point that the coming together of text and world can be appropriately colonized.

Thus, basing the spatial event on the notion of text in effect reinforces a double-disciplined literary geography, it also positions the concept within the confines of the modern distinction between "text" and "world" (see Chapter 1). Echoing the words of Braun (2004:169, also see Chapter 1), the spatial event is defined by Hones through her "proclivity" to "purify the world into essences" (in this case, "text" and "extra-text"), and only then, "become anxious about the results" of doing so (Braun, 2004:169). As we will see below, this is a problem that Thurgill and Lovell's notion of the hinge attempts to negotiate (2019).

The notion of the spatial event thus offers many opportunities for scholars working in literary geographies, as well as presenting them with a number of problems. This chapter suggests that, by being aware of, and openly engaging with, philosophical and conceptual innovations across the social sciences, arts, and humanities, the notion of the spatial event can be better contextualized and employed. For example, if we look beyond the limits of a double disciplinarity in order to integrate insights from philosophy, Alain Badiou's notion of the "event" can be informative (2005, 2006). In multiple writings, Badiou refined the common understanding of the event to refer not just to any spatial moment but one that has transformative potential – in other words, an event would bring something significantly "new to the world" (after Bassett, 2008:895). According to Badiou, an event is therefore an occurrence that realizes its innate potential to offer the exceptional, bringing about a "moment of rupture in time and space" (Bassett, 2008) and enduringly transforms the normal conditions under which life is lived. To this end, there are 4 interrelated aspects to a Badiourian event. Firstly, there must be an "irruptive" moment which brings something new to those experiencing it; secondly, this moment must offer transformative potential; thirdly, "fidelity" must be shown to the irruptive moment; and fourthly, the potential for transformation must be realized in practice. Due to the relative rarity of such phenomena, Badiourian events are commonly considered to occur on the scale of the dramatic, historical, and international (and include oft-cited exemplars of the Paris Commune and Christ's resurrection; see Bassett, 2008).

If we bring Badiou's notion of the event into conversation with literary geographies, we can see that reading literature could arguably be "eventful" for some readers; a book may "touch" their "life" (to quote Bradbury's Dr Faber, 2008:108, see also Chapters 7 and 13, for example) in such a way that it prompts reflection and recognition, and changes the way they see and act in the world. Informed by Badiou, reading fiction has the potential to breach the prevailing norms and orthodoxies of their life, it may offer a break which "is truly foundational" (Laclau, 2004:121), and may "change all the co-ordinates" of normality (Zizek, 2004:175). In this sense, read-literature as a Badiourian (spatial) event enhances our understanding of literature as transformative, at the scale of the individual reader, or indeed more widely.[9] Read-literature may transform our ontological reality, enchanting and re-integrating our modern understandings of the world. Such insights, emanating from beyond a double-disciplined literary geography, can only serve to enhance literary geographies and suggest that any narrow fixing of the subject is unnecessary or at least premature.

A second theory that has emerged to offer a practical vocabulary for a Honesian Literary Geography is Thurgill and Lovell's notion of the "spatial hinge" (2019). Thurgill and Lovell introduce their notion in the following way:

> Until recently, . . . there has been no term to describe the shifting of extra-literary perception in the unfolding of literary worlds in actual-world settings that are otherwise unassociated with texts and which occur as a result of the "spatial event". The . . . introduction of the "spatial hinge" accounts for this extension of literary space into extraliterary sites which feel to some readers as if they "share the same affective environment" as the text.
>
> (2019:16–17)

As we can see, the spatial hinge formally engages with the same territory as Hones, explicitly adopting its "modern" vocabulary and the subsequent heterogeneous associations necessary within the parameters of this iteration of literary geographies. In doing so, however, the spatial hinge also works back on the spatial event to implicitly clarify some of its aspects. As the authors outline,

> The "spatial hinge" names a process which extends a reading (and with it the text itself) into places previously unassociated with the text, which start to feel as if they belong to the text and, as a result, come to be experienced by

> readers as fundamental parts of its literary landscape, even where the author has made no connection to such a site in their writing.
>
> (Thurgill, 2021:153)

By identifying how the "extension of literary space" may be felt "by some readers", the notion of the spatial hinge directly implies that the "intra-" or "inter-textual" spaces connected through the "event" of reading are spaces predominantly or even exclusively intended by the text's author. The notion of the spatial hinge therefore implies not only that the author directs associations for the reader to spaces within the text, and to other texts, but also that there will be an accurate correspondence between the spaces that this event connects to. In raising these issues, the spatial hinge calls into question whether the notion of the spatial event allows for any spaces to be brought together that are either "felt" by the reader or "unintended" by the author. The spatial hinge's explicit inclusion of reader agency, beyond the power of the author to direct it, is therefore a positive extension to the notion of the spatial event (and as it does so, provides further useful opportunities for literary geographies to engage with practices of psycho-geography, see, for example, Pinder, 2001; Bonnett, 2009[10]). Yet the spatial hinge's emphasis on "extra-textual" spaces does appear to exclude the possibility for reader-generated "intra-" or "inter-textual" spaces being "hinged" together. Such possible exclusions seem to artificially preclude the potential of the reader's imagination in these modern categories.

Similar reflection is also possible with respect to what the "spatial hinge" actually refers to. As Doel suggests, in Thurgill's writings "there are at least five variants of the concept in play" (2023:279). Is the "hinge", for example, the original "space" noted in the text, is it the process of "extension" (to use the author's words) to a different space? Is it a "moment" (see McLaughlin, 2023) or a relational amalgam of all three? Could the hinge be a portal, a pivot, a convertor, or a transistor? If we were to take (one of) Saunders' interpretations of the notion, a hinge could be understood as a "moveable joint. . . , connecting two objects" (2023:253), and it is perhaps in this understanding that we see both the strengths, but also the weaknesses of the notion. A hinge both separates and connects, it precisely reflects Braun's modern proclivity, at once separating the text from the world, and subsequently seeking to connect them together.[11]

It is possible to argue, therefore, that the potential offered by the spatial hinge to establish and explore new literary geographies is inevitably limited by the modern constitution in which it is rooted, and the double disciplinarity in which it serves. We can see how this double disciplinarity "stifles" (after Sayer, 2003:2) what is claimed to be the "natural process" of this theory's development (Thurgill, 2023:234[12]) by again looking beyond a double disciplinarity and approaching the notion from cognate, but "othered" subject areas. In the interdisciplinary field of Modernist Studies,[13] for example, Miller (2013) explores the work of Lyotard (1990), who in turn examined Duchamp's essays on "transformations, partitions, machinations, and *hinges*" (Miller, 2013:93, emphasis added). From this work, it can be suggested that a "hinge" is part of broader transformative process: a metamorphosis from one "state" to "another", referring to the "operation of transforming one entity into another" and in particular, "the structural point . . . of transition and transformation between entities" (Miller, 2013:95). From this perspective, the hinge is not a space, or mechanism, or connector, it is the practice of changing something, as Miller puts it:

> by "general hinge," we understand a set of approximate operations permitting the dismantling of a work . . . and the reassembly of another . . . starting from the elements of the first one. A hinge between two states.
>
> (2013:95)

From this perspective, the hinge calls our attention to a process where one "entity" or "process" (what we may term assemblage to use the insights from Deleuze and Guattari, 2003; Delanda, 2006; Anderson and Saunders, 2016; Anderson, 2016b) is effectively dismantled, de-composed, or disassembled into its constituent components, and then reassembled, initially with the components of the earlier tracery, but also adding new ones. If we refer this insight back to Thurgill and Lovell's spatial hinge, these new components may be places "felt" by the reader to be congruent, coincidental, or somehow corresponding with a literary text, but not explicitly mentioned by the author. The emergent literary geography is a new "state" produced through the complex hinge-work operation of reading; a process we have seen in Chapter 1 to perform the "interactive" (to use Hone's term earlier), imaginative, and generative world of literary geographies.

Conclusion

Thus by looking beyond the disciplines of literary studies and human geography, it is possible to usefully align critical theories which give added depth to the notions of hinge and event. Indeed, the addition of these insights could be seen to show the benefit of a more diverse framing of literary geographies beyond a narrow double disciplinarity. By doing so, a different form of literary geographies can emerge which, in Pernecky et al.'s description of interdisciplinarity, lays beyond,

> conventional norms and processes of knowledge production, dissemination, and communication; [and] challenge[s] the established views about the scope and limits of what is possible, relevant, desirable, and even credible in academic terms.
>
> (Pernecky et al., 2016:390)

This book makes the case that it is only through literary geographies embracing a multi- or functional interdisciplinarity that all scholars interested in the complex can openly do so. It is important to respect the intellectual potential of these types of interdisciplinarity, but also offer the possibility of an intellectual space to those who may not sense their rootedness in the modern disciplinary landscape, and indeed lie beyond the confines of Hones' double disciplinarity. As Chapter 3 goes on to argue, if we recognize the significant changes in philosophical approaches and conceptual developments that have occurred over recent decades, a commitment to maintain modern disciplinary categories and a concomitantly narrow vision of academic endeavour is eroded. In such a context, we can recognize how scholars from an array of disciplines (including, for example, cartography, history, cultural studies, art, and geohumanities) who would be excluded from a Honesian literary geography have contributed to the possibilities emerging in literary geographies more broadly. Any version of our interdiscipline therefore ought to recognize this context, and the array of interests who wish to contribute to it further. Chapter 3 outlines this changing philosophical context, and the conceptual developments that have worked together to make literary geographies such a diverse and plural subject area. From this, the chapter offers its own definition of the subject, suggesting that the contemporary moment offers the opportunity to reframe disciplinary debates and (b)orders, offering templates and directions to enable scholars to realize the affinities felt beyond their own disciplinary "towers", and cast off from them into new "waters" of academic interaction.

Notes

1 This accusation is perhaps particularly relevant with to human geography as discipline, with its history of colonization in both an academic and political sense (see Driver, 2000; Anderson, 2021).

2 Whilst also referring to the act of highwire walking between the original towers of the World Trade Centre which is the focus of her key subject in Let the Great World Spin (2014).

3 Hones does not specify here what type of geography she is referring to here; one might assume she is referring to human geography (as she does elsewhere in her work), presuming that for the majority of her readers geography as a category is enough to establish her key point, and from this the "space between" is where literary geographies emerge. But in doing so, she perhaps inadvertently masks the tensions in the relations between human and physical geography, and within each of these groupings the broad diversity of approaches within each.

4 Hones does draw on Soja later in her 2014 work when engaging with the betwixt and betweenness of the "real" and "imagined" conjured through read-literature (2014:66).

5 In this vision of the interdiscipline, it also remains unclear how the "towers" of literary studies and geography can respond to changes and "turns" (or what Chapter 3 terms "currents") in philosophy or conceptual development. Can fixed towers with sturdy foundations "turn"? Do their occupants rotate their gaze to new horizons? Can towers, and their occupants, be moved when philosophical changes require it (across the land, for example, or politically, or affectively)? Perhaps, like cities in Reeve's *Mortal Engines* (2018), the endgame for such disciplinary towers is to become mobile and trundle across disciplinary landscapes, imperialistically (after Sayer, 2003) enclosing other disciplines.

6 Indeed, Doel went on to suggest that seeking any fixed territory for literary geographies may even prove to be a pointless exercise, suggesting that

> Neither [disciplinary] space nor spacing will ever [be] staked out and pinned/penned down, least of all by a taxonomic or taxidermic operation. Space – cut – slips away. No matter how rigidly and tightly the partitions and the seams are drawn, space and spacing will have spilt out and leaked.
>
> (2018:45)

7 Furthermore, the homogeneous nature of these key disciplines remains the subject of contentious debate. As Johnston alludes to earlier (and see, for example, Hennessy and Means (2015)), is human geography already an "interdiscipline"? If so, what does this mean for the nature of what a "double-disciplined" literary geography definitionally is?
8 In this way, the spatial event complements moves within literary geographies to develop notions of assemblage emergent from critical studies and philosophy (see, for example, Anderson and Saunders, 2016).
9 For the latter, one could suggest literary traces such as Orwell's 1984 (1949), Carson's Silent Spring (1962), Abbey's Monkey Wrench Gang (1975), or Conan Doyle's Sherlock Holmes (1903), as spatial events in the Badiourian sense which have offered sustained and prolonged transformation to a range of literary geographies.
10 Psychogeography, as Bonnett (2009) recounts, derives from Lettrist and Situationist International practices in the 1950s which sought to reimagine everyday space through new forms of interaction. More recently, there has been a revival in psychogeography across more traditional and literary circles (see, for example, Ackroyd, 2000, 2007; Ho, 2006; Sinclair, 2003; see also; Moorcock, 1988), and for literary geographies specifically, it is possible for interactions between a book, reader, and location to alter the "physical and psychological landscape of the city" (Pinder, 2005:386) for an individual. Through escaping with characters into places, it is possible for a novel to have a psychogeographical effect on a reader by helping them to "re-imagine the forgotten nooks and crannies of ordinary landscapes [and] re-enchant and remythologize prosaic geographies" (Bonnett, 2009:46). The spatial hinge suggests that these places may be re-imagined by readers whether they are directly referenced within literary texts, or not, and become part of an actor-centred geographical assemblage of literary geographies.
11 Thurgill appears to reluctantly acknowledge the bicameral confinement this approach creates when he reflects on his work and acknowledges: "readers encounter textual geographies not through a process of spatial bifurcation, but through the unification of spatial realms, an inhabiting of the all at once social-spatial-textual" (2023:262).
12 It remains important to note that the development of this theory is not "natural" in nature, but as this chapter has contended, part of the politicized development of a particular interpretation of literary geography (see also Chapter 3).
13 Often referred to as the studies of literature, psychoanalysis, art, music, film, and architecture, see British Association for Modernist Studies (no date).

3 New literary geographies

A confluence of approaches and disciplines

As this book has argued thus far, scholars from a range of modern disciplines have come together to redefine and recompose the world of literary geographies. As this has occurred, competing claims have arisen over how the emergent interdiscipline should be organized and controlled, with various interpretations gaining support and critique. These contemporary "turf wars" are not necessarily a new phenomenon, as attempting to frame the subject of literary geographies, in all their plurality and diversity, has never been an easy task.

Literary geographies have always explored a range of different questions, been motivated by different onto-epistemologies, and harnessed a range of methodologies. However, working with the confines of modern disciplinary categories has discouraged scholars from recognizing and forging alliances between cognate work in separated areas.[1] Up until recent years, therefore, an impression was given that scholars working within literary geographies did so within the limits of their originating disciplines, writing to and for those who share their philosophical dispositions and methodological orientations.[2]

Yet, as Nissani points out, the modern model for categorizing and disciplining academic activity is "inaccurate . . . and misleading" (1997:202). Mimicking the modern process of atomization and separation (see Chapter 1), this model has often had the effect of dividing activity into "numerous blocks of specialization" (1997), rather than emphasizing connections and synergies. Remaining resolutely organized in terms of terrestrial metaphor (we may speak, for example, of the disciplinary "landscape", "fields" of expertise, and "turns" into new "territories"), this model normalizes a "sedentary" metaphysics (after Mallki, 1992; see also Cresswell, 2004) and encourages academic activity to be considered as durable and isolated (rather than, for example, "precarious" and "relational", after Philo, 2005).[3] As a consequence, as Nissani suggests, there is a

> need to reconceptualize our model of disciplinary growth and specialization, adopting a more organic model that accounts for the intricate links among the many specializations.
>
> (1997:202)

DOI: 10.4324/9781003321248-3

From this beginning, this chapter casts off terrestrial metaphor to adopt a more fluid vocabulary to story academic activity. Indeed, in recent years, a number of scholars have been mobilized by metaphors of water to reframe their approach to the world (see Anderson and Peters, 2014a; Brown and Humberstone, 2015; Peters and Steinberg, 2015; Peters et al., 2018[4]). Reflecting its obvious liquid materiality, this exploration has generated a "fluid ontology" which frames the world as a circulation of flows, immanence, and transience, rather than any fixed, fossilized, durability (see Peters and Steinberg, 2015).[5] Buoyed by these insights, this chapter will adopt the idea of "sea change" to refer to the widespread influence of a fluid, relational ontology mobilizing the social sciences, arts, and humanities, and the notion of "currents" to capture the change of emphasis in conceptual and methodological activity that this sea change has set in motion. In particular, the chapter will identify 4 key currents which have come to influence an array of traditional disciplines: the current of culture, the current of space, the current of cartography, and the current of creativity.[6] The chapter adopts this new model in order to " 'jar' against accepted ways of thinking, speaking, and writing" about academic activity, and as a consequence, "opens up possibilities for dialogue" (Noxolo et al., 2008:147) about how these sea changes and currents have had unique significance for literary geographies.

Indeed, this chapter suggests that due to these sea changes and currents, literary geographies is now operating in perfect storm conditions for innovation and influence. Whilst the currents have mobilized a range of scholars, the former's particular significance for the component dimensions of literary geographies has led to new voices contributing to the subject area. As a consequence, more scholars are now interested in culture and literature than those operating solely in the disciplines of literary studies, cultural studies, or cultural geography. Similarly, many more scholars are embarking on "spatial adventures" (after Casto Broto and Baker, cited in Bridge, 2018:1) than those operating solely in human geography. These changes have enticed new interest in the processes and practices of mapping (thus expanding the disciplinary frontiers of cartography), and attracted many scholars to experiment with creativity in method and academic output than ever before. These combined energies have thus, as Thacker (2017) notes, created many new routes (or "paths") to literary geographies, offering the opportunity for once-separate disciplines to be brought together in ways that have yet to be fully grasped or understood. In this context, the final section of the chapter will provide an open definition for literary geographies, a definition which seeks to enable all of these growing interests to find an intellectual space of operation and alliance. This definition will also set the platform from which the *Literary Atlas* project's own plotted approach can cast off, and this approach will be outlined in Chapter 4.

A sea change towards relations

Over recent decades, a sea change has been felt across the social sciences, arts, and humanities. This sea change has moved scholarly attention from a fixation on the world as constituted by the modern, and towards an interest in fluidity and relationality. The modern constitution, according to Latour (1993), emanated from the attempt by Renaissance scholars to grapple with and control the "multitudinous, churning, spewing energies" of the world (to use Philo's words, 2005:826). This was achieved by using empiricist methods to identify, itemize, and give names to what they considered to be discrete aspects of existence. The process replaced a chaotic amodern reality with a mechanistic ontology of stable forms; a reality where "things" were reified into significance, context excised from agency, and order constructed in the place of uncertainty. This structuring, understanding, and controlling of the world became the foundation upon which modern science and anglophone-influenced academia were based[7]; indeed for many this modern constitution remains so successful that it, along with the discrete "things" it brought into being, has long been taken for granted. For the majority, the reality of the modern constitution has become doxa, the only way the world can be.

> If there is anything that makes us truly "modern", it is our proclivity, first, to simultaneously purify the world into essences all the while furiously producing ever new heterogeneous associations, and, second, to only subsequently become anxious about the results.
>
> (Braun, 2004:169)

As Braun suggests (earlier, see also Chapter 1), for many years, scholars have become "anxious" about the modern move to "purify" existence into "essences". Indeed, over recent decades, a host of scholars from a range of disciplines have argued for an amodern "relational" or "fluid" ontology (see, for example, Doel, 1999; Bauman, 2000; Murdoch, 2005; Haraway, 2003; Steinberg and Peters, 2015). This move has been propelled by those who, despite acknowledging the practical utility of the modern atomizing process (simplifying processes so they can be communicated and engaged with), nevertheless sense that it does not completely capture the world as it is. For these scholars, the creation of the modern atomized "unit" has imposed stasis upon entities and, through this process, a degree of vitality has been lost. For them, how modern "things" change and evolve, and how this change involves combination with other "things" over time, is crucial to their understanding. Thus, from this perspective, the imposition of fixity and independence on "things" is dismantled and replaced with an interdependent, constitutive ontology, where "things" are always redefined through their coming together with everything else.[8]

A fluid, dynamic ontology thus abandons the isolated agency of the modern world and (re)imposes a perspective where relational agency holds sway. Here, the success of one "unit" is structured by the capacities of other "units" within the relational complex (after Rouse, 1996), the network may effect a greater capacity through synergistic combinations between its actants, or its capacity may be negated through resistance or annulment within the "network". A relational ontology therefore provides an important change of emphasis when compared to the modern constitution. We may remain agnostic about whether we start with the notion of discrete "things" because that is how they are, or because that is how we have installed, named, and normalized them; but now they are nevertheless brought into interdependent relation. As a consequence, we/they/things are now hybrids, consisting of flows and connections within which these are continuously (re)constituted (Braun, 2004:171). This represents a shift from a world of ontological stability (however manufactured and imposed) to a world of ontological instability. Instead of inhabiting a world where clear ontological entities act on other clearly categorized entities, the world is now composed of flows and connections – of "beings" that are also "becomings" (to use the words of Whatmore, 2002), of "entities" that are also "entities/processes" (to use the words of Harvey, 1996), and of "noun chunks" that are really "verbs" (to use the words of Laurier and Philo, 1999). This sea change has opened up new opportunities that problematize but also offer possibilities to a range of scholars to reassess their conceptual engagement with the world. This chapter moves on to document 4 key currents of engagement that have been set in motion by the sea change towards fluid relationality: the currents of culture, space, cartography, and creativity.

The current of culture

This book contends that the sea change towards relationality can be seen to resonate with, and further propel, a range of other currents which influence contemporary literary geographies. The first of these currents worked to fundamentally reboot how the notion of "culture" (including the medium of literature) was approached and understood. Known in modern vernacular as the cultural "turn", in the 1970s and 1980s, this current eroded the assumption that culture was abstract and artefactual in nature. Buoyed by the sea change towards relationality, and propelled by the work of scholars such as Foucault (1973, 1980, 1984, 1995), Barthes (1977), Deleuze (1985), and Deleuze and Guattari (2003), it was suggested that cultural products did not exist in isolation, but could (only) be understood as part of a broader set of ideological and political relations. Cultural products were deemed to represent power and meaning, and if these meanings could be "detected" (after Turkel, 2006:259), then culture could help explain the structuring of human relations.[9]

The recognition of culture's symbolic and political value led to a revitalization of many subject areas.[10] In terms of the subject of this book, this re-contextualization had implications for the framing of literature. Due to this current, there was a move away from considering literature as a "documentary" source (after Brosseau, 1994:333) which could be used "unproblematically" (see Sharp, 2000:327). No longer could literature be deemed to function as a faithful reflection of a straightforward world – as this neglected the ideological discourse which inevitably framed the positioning, authorial voice, and overall production of a literary text. The situatedness (after Haraway, 1988) of "fictional knowledge" was no longer deemed to offer a transcendent encapsulation of a person, incident, or location, rather it was acknowledged that literary fictions offered one way of viewing the world, positioned as a particular view from a particular somewhere (and constituently influenced by received genre, authorial intent, and subjective agency).

However, as we have noted in Chapter 1, literature is not simply defined by the crafted words that are written but also by how they are read and subsequently interpreted. As such, the current of culture re-installed not only the situated author into the network of relations that produced literature but also the situated reader (see Barthes, 1977). This recognition emphasized that a book, like any cultural text, was never simply a "stable ontic entity" (after Kitchin and Dodge, 2007:341, cited in Hornsey, 2012:675), but rather a "fundamentally ontogenetic" process, created "through the moment of [its] use".[11] From this perspective, read-books became dynamic processes, only realized into being when an author, publisher, and audience come together to "make it happen".[12]

The current of culture can thus be seen to shift academic understanding of human products from an emphasis on "artefact" to an interest in "activity"; the current of culture brought "contexts" back to "texts". As such, this current offered particular significance for literary geographies; it revitalized the possibilities of studying literature and, as a consequence, led many more scholars (including human geographers and scholars working in cultural studies) to become engaged with literature as a source material. Indeed, interest in this current gained further momentum as it combined with the second key outcome of the sea change towards relationality: the current of space.

The current of space

The "spatial" current which swept across the social sciences, arts, and humanities involved a renewed recognition of the importance of *geographical* context to human life (see Cook et al., 2000; Lossau and Lippuner, 2004; Martin, 1999; Warf and Arias, 2008). Driven in part by globalization and the increased connection between places, industries, and environmental flows, the spatial current involved a growing acknowledgement that "space" and "place" were a key influence on identity, community, the economy, and decision-making. Contemporary issues were no longer framed as abstract processes occurring in a vacuum, but rather as positioned activities co-produced by (and co-producing) a complex of "geographical" factors. As a consequence of the spatial current, geographies were rarely framed in a narrow "locational" sense, but rather activated to suggest a range of other social processes. Physical places were deemed to combine temporal, social, political, *and* cultural in their nature (see, for example, Soja, 1996; Cresswell, 1996) – specific geographies became the medium in which a range of political processes were made manifest.

The (re)discovery of the importance of geographical context led to the study of place changing from one that was seen to be "highly abstract and remote from experience" (Tuan, 1975:151) towards one that considers place as "thoroughly meshed" (Casey, 2001:684) into all aspects of the human condition.[13] Indeed, with the spatial current mobilizing the social sciences, arts, and humanities, not only was geography deemed to be central to human life, it was also considered in new ways. Influenced by the relational sea change, many scholars no longer framed geographies as solidified and singular, but rather dynamic and multiple (see Massey, 2005, for a full review). Where once scholars would contend that it is every geographical site's "persistent sameness and unity which allows [it] to be differentiated from others" (Relph, 1976:45), relational places were now seen to be "always in a process of dynamic unfolding and becoming" (Rose, 2002:385).[14]

As a consequence, new notions of place and space became vital to reflect on and integrate into a range of inquiries. With respect to literature, this was undertaken with particular effectiveness by Hones (2014). Without directly acknowledging its antecedents, Hones built on the recognition that a book should no longer be seen to be part of the "object world" but rather as part of the "event world" (see Ong, 1971 and Pocock, 1989, for example, earlier), and explicitly installed "spatial" into this emergent constitution. For Hones, a book became not simply as an event but a "spatial event" (2014, see also Chapter 2). Through installing this context, the influence of the geographical current is fully acknowledged; it is clear that any book is a co-combined geographical, temporal, and social "happening", activated by reading. In Hones' words, book should be seen as "events that happen in the course of sociospatial and intertextual interactions" (2014:11), and "understood as an inherently spatial practice" (2014:23).

From this example, it can be seen that the cultural and spatial currents have had a particular synergy for the subject of literary geographies, fundamentally influencing how two of its key aspects, literature and place, can be considered relationally. But as these currents rendered our understandings of culture and geography more complex, questions arose in terms of how it may be possible to chart, capture, or otherwise "map" these newly configured aspects, and most importantly, their relations. It is with respect to these questions, that the third key current made waves: the current of cartography.

The current of cartography

> Mapping, it seems, is [now] a universal conduit for thinking about and through a wide, diverse terrain of topics and issues.
>
> (Kitchin, 2016:7)

With the spatial current mobilizing many subject areas, interest in the formal discipline of depicting geographies – cartography – has concomitantly risen in recent years. According to Kitchin, to map is "to set out places, ideas, events, findings, etc., in relation to one another – [and] is a powerful means of sense making" (2016:6). Indeed, cartography itself can be seen to reflect and synergize with the relational changes exemplified earlier. Where once cartographic representation sought to exhibit the philosophical conventions of the modern, "purport[ing] to be 'natural'" and producing "objective visual representations arising out of scientific observation" (Owen, 2002:10), with the rise in the importance of relational thinking, it is now deemed "too easy" to frame cartography in such a "naïve" way (Kitchin et al., 2011:2).

In the context of the sea change noted earlier, it is now considered that, like any other cultural or academic intervention, maps and the act of mapping are defined by contexts which determine their focus and framing (most notably in the case of cartographic design, acts of scoping, projection, orientation, symbolic representation, and scale). As a consequence of these political acts, cartographies are now widely perceived as "social constructions" which are "products of power and its exercise" (Kitchin et al., 2009:9); they are "a set of practices rather than simply representations" (Kitchin, 2016:6), situated and contextual "arguments" (Wood, 2015:306) rather than "artefacts".

The influence of the cultural and spatial currents can therefore be seen on cartography, and in the words of Chavinier et al. (2016:83), they present a number of specific "challenges" which have energized the practice of map-making. As we have seen earlier, the cultural current now requires us to consider maps as produced by situated human authors (see, for example, Wright, 1942; Rees, 1980; Harley, 1988; Monmonier, 2007), and engaged with in different ways by a variety of map "readers". Indeed, as maps are now political arguments rather than artefacts, questions emerge over whose views are being represented, whose voices are being heard, and whose worlds are being brought into being through the cartographic process. Concomitantly, the spatial current requires us to consider places as emergent, dynamic, and always changing. A further plurality is thus set in motion, as situated perspectives of the same location are infinitely multiplied moment by moment. Such complexity fundamentally challenges the abstracted, atomistic, and durability of the modern cartographic imaginary, as Chavinier et al. put it:

> How can Euclidean geometry be made to coincide with other types of spatial representations linked to movement and rhythm? . . ., how can maps express . . . temporalities? . . . What happens when each individual becomes an actor, in particular an undisputable spatial actor?
>
> (2016:83)

The cartographic current is therefore charged with these questions: how to map the implications of relational, cultural, and spatial turbulence? Due to the influence of energies from the cultural and spatial current, a growing diversity of scholars have begun to respond, using their own insights to challenge what we might consider maps to be, how they may be read, and what they may be used for. In doing so, they have set in motion another current of academic activity: the current of creativity.

The current of creativity

The effects of the cultural, spatial, and cartographic currents have led many scholars to become creative in the ways they reflect and communicate an amodern world. Although, as Delyser and Hawkins (2014) note, all traditional academic outputs (including peer-reviewed journal articles, textbooks, and monographs) are creative in their nature, many scholars are less interested in solely conforming to modern conventions and set out to articulate their insights in new ways (see Hawkins, 2011, 2012, 2021). This may involve directing their work towards different stakeholders (for example, wider publics rather than traditional academic colleagues), or producing new forms of cultural output (for example, artworks in a more general sense, rather than simply maps, papers, or monographs); as Hawkins sums up, scholars are increasingly seeking to engage with

> a wider field of creative practices, including critical-creative writing practices [or] a suite of visual culture methods, videography, photography and comic-book development.
>
> (Hawkins, 2012:64)

In sum, the creative current has led scholars to collaborate "with artists and even become creative practitioners in their own right" (Hawkins, 2011:464).

Indeed, the will to create has synergized with a growing demand from audiences, as Denison and Markula (2003:17) state,

> As readers become more accustomed to new types of narrative studies, the demands for sophistication, intelligence, and competence will only rise. Instead of writing just any old story, scholars "will be expected to [produce] critically informed stories with the power to illustrate, illuminate, inspire, and mobilize readers to think and act critically and reflexively" (Denzin, 1998).
>
> (2003:17)

The creative current has thus led scholars to consider afresh the communicative means through which they articulate their questions, methods, and understandings. Critical reflection has occurred with respect to what different languages of communication can do: in terms of providing appropriate questions, creative methodologies, and stimulating outcomes, in terms of the scale of audiences authors and scholars wish to engage with, and what sort of relation they wish to invite. This reflection has coincided with, and further propelled, innovation in two key areas for creative mapping: *technologies* and *techniques*.

The current of creativity: technological innovation

Technological innovations have been embraced by many scholars "to cope with the complexity of [not only] literary space" (Piatti and Hurni, 2011:221) but the relational complexity of the spatial current in general. A range of innovations in digital technologies and software, including geographical information systems (GIS), modelling capacities, visualization techniques, animation, and positioning systems, along with virtual and augmented digitization mean that many scholars have been able to explore,

> how technologies can be used to manage, manipulate, visualize and analyze the geographies embedded within texts of all kinds.
>
> (Gregory and Cooper, 2013:267)

For scholars working in literary geographies, this application of technological innovation has, as Lilley identifies, "tended to follow a social science model of research, geared towards mapping and analyzing quantitative spatial data-sets" (2011:26). In the main, this has involved replicating Moretti's "distant reading" of literary works (see Moretti, 2005) – a distinctly modern practice of mapping which we might understand as involving "abstraction", "reduction", and "extraction". Distant reading involves scholars "abstracting" themselves as readers and cartographers from the understandings being produced,[15] then "unapologetically eschew[ing] detailed textual engagement" (Gregory and Cooper, 2013:270). This is achieved through "reducing" the crafted text to specific words or phrases, and then these items are "extracted" from their context, codified, and visually displayed. In Franco Moretti's words, to conduct this process:

> You choose a unit [from the fiction] – walks, lawsuits, luxury goods, whatever – find its occurrences, [and] place them in space . . . in other words: you *reduce* the text to a few elements, and *abstract* them from the narrative flow, and construct a new, *artificial* object.
>
> (2005:53, emphasis in original)

Mimicking the modern atomizing process, the "abstract, reduce, extract" formula seeks to not only minimize the "uncertainty" and ambiguity of the text (Reuschel et al., 2013) but also enable simplified "data spaces" to emerge from it (Benedikt, 1991:135). Through this process, a literary cartography is produced in the form of "a map [that] is . . . geometric figure" (Tally, 2013:46), which (for better or worse) may not offer any strong resemblance to the literary work itself.

The current of creativity: innovation in (writing) technique

For many scholars influenced by the relational sea change, innovation in technique has become more important than the modern adoption of technology. For these scholars, the cartographic current prompted them to reconsider the capacity of the crafted word. It was suggested that literature retains the potential to conjure a "textual cartography" (Roberts, 2016:2), with words "aspiring to yield what a conventional [i.e. modern] map . . . cannot even come close to conveying" (Roberts, 2016). Such conveyance may involve producing "a map attentive to the experience of place" (Wood, 2015:307). This mapping process effectively seeks to counter the "abstract, reduce, extract" formula by positioning writers as "being in the world" they are describing (in the phenomenological tradition, after Merleau-Ponty, 1962). Such mapping, claimed by writers such as Least Heat-Moon (1999) and Naramore Mayer (2014), could also perhaps be associated with any writing that explicitly seeks to project a location in "depth". By rarely using graphic or illustrated images in the modern convention, and focusing instead on the capacities of crafted language to communicate an experiential, layered, and immersive account, these works inevitably raise the question of "what is meant and understood by the term 'map'" (Roberts, 2016:2). In response, many consider read-literature to have what we may term as an intrinsic cartographic capacity, as Tally confirms,

> In a manner of speaking, literature . . . functions as a form of mapping, offering its readers descriptions of places, situating them in a kind of imaginary space, and providing points of reference by which they can orient themselves and understand the world in which they live.
>
> (2013:2)[16]

This challenge to the very idea of what a map is or could be has opened up capacities for technical and technological innovation to come together; scholars have responded to the cartographic current by going deep. Deep mapping is a rich and textured practice (see Pearson and Shank, 2001; Bodenhamer et al., 2015; McLucas, 2024), but can be seen to emerge from the literary tradition which sought to offer an experiential, layered, and immersive account of the world. Deep mapping seeks to include time, history, and lived inhabitation of a location, to create multiple perspectives which seek to offer a less "artificial" (see Moretti, 2005), and perhaps more "authentic" sensation of a location. Echoing anthropological impulses to go beyond "thin" descriptions (which produce thin maps (after Harris, 2015) in preference for "thick descriptions" (after Broyard, in Least Heat-Moon, 1999:8)), deep mapping,

> means observing the details of life until they begin to coagulate or cohere in an interpretation . . . Apart from sheer sensuous pleasure, its gives you the comforting feeling that you're not altogether adrift.
>
> (Broyard, in Least Heat-Moon, 1999:8)

Thus deep mapping seeks to produce "a cartography of depth. A diving within" (Roberts, 2016:6), perhaps even a "deep geography" (Harris, 2015). In doing so, this process may offer an illusion to something more "profound" (Naramore Maher, 2014:10) than the thin modern object, and could be seen to be "radically nostalgic" (to use criticism of resonant psychogeographical approaches, see Bonnett, 2009) in seeking to discover the quintessence (or genus loci) of a place.[17] Through practitioners such as McLucas (2024), however, deep mapping has avoided such sedentary pitfalls and has embraced the plurality of narratives about a place; in this interpretation, deep mapping involves a range of data sources and types which offer not an authoritative or singular account, but often contradictory positioned accounts. As Robertson and Mullen state,

> The adjective deep . . . expresses the concept of multiplicity. The aspiration of deep maps as captured in its founding definitions is to capture a multiplicity of sources used, of perspectives represented, of experiences captured, of interfaces used within the framework of the deep map. This variety of inputs to the deep map [have become] central to the concept. It distinguishes

> the deep map from other kinds of . . . approaches, especially computational methods, which too often rely on a single kind of source such as documents in text analysis.
>
> (2022:132)[18]

The inclusion and juxtaposition of multiple sources transform mapping from being a pickled artefact into a processual archive. No longer are maps reduced to being a "false staticization of social processes" (after Steinberg, 2014:xv) but become "densely layered, thickly embedded, and highly contingent environment[s] that allows us to explore the development of place in all its spatiotemporal complexity" (Bodenhamer, 2021:7). The current of creativity, with all its technological and technique-oriented dimensions, therefore offers scholars a range of solutions to engage with the consequences of the cultural, spatial, and cartographical currents that are washing through the arts, humanities, and social sciences. These combinatorial energies offer huge significance for literary geographies, attracting new attention and offering new opportunities to map these worlds differently.

A perfect storm of literary geographies

As we have seen, a new model can be used to understand the ways in which academics not only approach the world but also operate within it. By adopting more aquatic metaphors, in this case towards sea changes and currents, we can dissolve existing lexicons to view afresh how relationality has effected a range of disciplines, and set in motion a number of new conceptual developments. These changes have substantively influenced how scholars approach the world, but also encouraged them to look beyond the limits of their home disciplines and recognize the cognate work occurring in other areas. As a result, many more scholars have become interested in culture and literature than those operating solely in the disciplines of literary studies, cultural studies, or cultural geography, and more scholars are now interested in place and spatiality than those operating solely in the discipline of human geography. These changes have enticed more scholars to explore and critique the processes and practices of mapping (and thus proliferating interest beyond the conventional discipline of cartography), and attracted many scholars to experiment with creativity (and thus broadened interest beyond the purview of art scholars). Indeed, one could suggest that the diversity of scholars working in literary geographies may now be "housed" in a range of disciplinary domains, including, for example, the creative geohumanities, history, mobility studies, literary studies, discourse analysis, human geography, imaginary geographies, reception studies, cultural studies, creative tourism, (cognitive and/or corpus) linguistics, narratology, heritage studies, textualism, tourism, print cultures, geocriticism, geopoetics, cartography, topology, stylistics, art criticism, sound studies, geospatial information science, and spatial literary studies. These combined energies have thus brought conventional disciplines together in ways that have yet to be fully grasped or understood, yet may come to exemplify a process which has significance across the academe; namely, a growing move away from understanding scholarly disciplines as singular, coherent, and uniform, and towards a recognition that established structures are better defined by "quasi-independent communities", populated by scholars who experience "a substantial degree of isolation from others within the[ir] discipline" and enjoy "stronger links to others [outside]" (Johnston, 2000:974).

Scholars have thus cast off into these new waters of academic activity from a range of different directions and with a range of different trajectories. Some may have started out explicitly identifying with the disciplinary label of "literary geographer", but many may not have (they may feel more at ease, for example, with the label of literary critic, or data analyst, or creative mapper). Each will be buoyed to their positioning by the onto-epistemologies, philosophical orientations, theories, and methods of their origins, but are nevertheless subject to the possibilities offered by the turbulent present in which they operate. In this context, this chapter concludes by setting out a working definition that seeks to harness the reciprocal influence scholars now have not only for the future of the subject but also for the disciplines from which their journeys set out. To that beginning, a definition of literary geographies is offered which provides the basis for this future, as well as the mooring from which the *Literary Atlas* project's own plotted approach will set out, and will be introduced in Chapter 4.

New literary geographies: towards a functional interdiscipline

The following working definition of literary geographies emerges from and seeks to be sensitive to the sea change towards relationality as well as the combined currents of culture, space, cartography, and creativity that have been discussed earlier. It is not intended to actively prescribe or limit the subject's (b)orders, indeed it is hoped that those scholars working beyond or before these energies can operate effectively within it.[19] This book therefore offers the following working definition of literary geographies:

> *Scholars working in literary geographies will critically produce geographical readings of fiction, with a view to stimulate new understandings of the agency of this complex on (the cultural, spatial, cartographic, and creative nature of) the human condition.*

Drawing on this working definition, 3 key components can be seen to define literary geographies. Firstly, scholars working in this subject area will have a specific focus on literary fiction. This literature may come in a range of forms, for example, novels, short stories, essays, poems, or plays (and variations thereof). From this central concern, literary geographies have the potential to proliferate: literary geographies may expand to have a concern for adjacent performative categories to literary fiction (e.g. audiobooks). Perhaps more tangentially, literary geographies may come to have a focus on an interpretation of this literature through theatre or film direction. With these possibilities in mind, it is prudent to distinguish why the "root" of this tracery matters: in other words, why does a literary geographic line of enquiry produce a particularly valuable insight from its efforts?

A key answer to this question is provided by the second key distinguishing feature of this working definition: scholars working in this subject area are committed to producing explicitly geographical reading (of whatever type) of literature (or adjacent cultural form). It is clear that, following the spatial current, it is possible to argue that all or any readings of literature would inevitably be spatial to some degree, but working in the subject area of literary geographies would suggest an explicit acknowledgement of this position from the outset, and a commitment to argue that this geographical reading (of whatever type) offers a crucial insight into the nature of the worlds that emerge through it.[20] Such an approach leaves the complexity and diversity of interests of a specific project to be made transparent; it does not mean that all scholars working in literary geographies should be interested in the straightforward location in which stories are set – those who are interested in history, or gender, or ethnicity, or sexuality of characters, or dramatic arc, or the language crafted to compose them, could and should do so. However, integral to these readings should be the explicit acknowledgement that spatiality in its broadest sense is a key dimension to each of these lines of inquiry. In a similar vein, it is important to note that a geographical reading is not essentialist in nature – geographical reading can occur in a range of forms (from distant, deep, close, plotted, etc.), with each demanding different methods and techniques. The integral "DNA" which forms their familial resemblance as "literary geographical" is the interest in exploring the spatial nature, influence, form, emergence, and consequence of the literature through this method.

The third and final core component of the working definition is the understanding that scholars working in literary geographies need to be critical in their work. Following the combinatorial currents outlined, scholars may be artistic, they may be curatorial, they may produce monographs, conventional maps, virtual reality environments, sonnets, walking tours, or even merchandise (or other yet-to-be-imagined outcomes), but these outcomes are required to be critical in nature – in other words, they must be contextualized and positioned within the panoply of academic, societal, and cultural work which these outcomes respond to and engage with. Work emergent from an academic associated with literary geographies would therefore be different from work originating from an artistic, journalistic, or political position (with their own conventions, codes, and motivations). Based on these principles, scholars working in literary geographies have a platform to share the same working space as others, whilst also sharing a commitment to be open, transparent, and accountable with respect to their objectives, trajectory, methods, outcomes, and ends. It is from this basis that Chapter 4 moves on to outline how the *Literary Atlas* "plotted" its own course through the groundswell of interest in literary geographies.

Notes

1 Due to the prevalence of strict disciplinary approaches, attempts to chronicle the subject's history, for example, have tended to focus on the perspective of a particular discipline, or even sub-discipline. Bradshaw and Brown (2018), for example, offer a "little

history of literary geography" from the perspective of human geography, whilst Brosseau (2017) and Johnson (2004) offer valuable insights from the perspective of cultural geography (and similar projects could be undertaken for literary studies, cartography, history, heritage studies, art, etc.).

2 As Hones puts it, "until quite recently most works in literary/geographical studies have had discipline-specific projected readerships and citation horizons" (2014:1).

3 Furthermore, as we have seen in Chapter 2, such nomenclature may predispose us to consider academic change in terms of "expansion", "colonization", and even "annexation", with unfortunate echoes of academic progress being equivalent to a military campaign.

4 In a range of subject areas, the physical materiality of water worlds has moved from the margins of consciousness to the mainstream. As this move has occurred, the consequences of "thinking *from* the water" (see Anderson and Peters, 2014a; Brown and Humberstone, 2015) have also been explored.

5 This ontology has discouraged the "imposition of clear, stable ontological categories onto the world" (Anderson and Peters, 2014a:10), and recognized that any categories should be seen as processes that have "only temporarily stabilized" (Anderson and Peters, 2014a, see also lisahunter, 2019).

6 In the modern nomenclature, "turns" towards culture, space, cartography, and creativity have been widely discussed. As such, for many, these movements have become normalized, but for others, they may remain unacknowledged or up for debate. It is necessary to briefly discuss these "currents" here in order to identify their existence and recognize their synergistic influence over a range of subjects, in particular literary geographies.

7 Amodern reality was thus converted into an ordered and atomized "modern reality", whilst also detaching the scientific observer from that world. Knowledge was itself divided into that which could be cognitively processed by the rational mind, and that which was sensed by the emotional body. Modern knowledge was thus categorized as abstracted, intellectual, reasoned, and cognitive, with the embodied affects of passion and emotion deemed irrational and thus silenced as "nonknowledge" (Pile, 1994:259). As we will see, this remains an issue with consequence for how literary geographies are framed and understood.

8 Many theorists have conjured ways to mark a shift away from the independent conceptual categories of the modern constitution. In Haraway's terms, it is the "relations" between modern "things" that should be "the subject of our analysis" (2003); as Braun puts it (earlier), we should be more interested in the "heterogeneous associations" than the acts and consequences of purification. Interest is therefore focused on the identification of relational complexes (Rouse, 1996), "hybrids" or "quasi-objects" (Callon, 1986:211), driven by the recognition that "in the real world everything is continuously 'interfered with'" (Lewis, 1960:45) and co-constitution dominates what "things" are at any point in time. In aggregate, these approaches accept that definition is always distributed and meaning is always molecular.

9 Scholars such as Barthes (1977), for example, framed culture as "text", some-"thing" that emanated from relations of power. In his view, texts attempted to convey messages with meaning and sought to perpetuate or challenge existing societal relations. From this perspective, culture was evidence of, and with, agency. What scholars had to do was "read" these "texts" and attempt to diagnose, interpret, or decipher what they represented; in the words of Aitkin, many scholars " 'read' images such as paintings, maps and landscapes, along with social, cultural, economic and political institutions as texts" (1997:198). Reading cultural texts was,

> the kind of evidence that [scholars could] now use for interpreting the symbolism of cultural landscapes . . . paintings, poems, novels, folk tales, music, film and song [could now] provide as firm a handle on the meanings that places and landscapes possess, express and evoke as do more conventional 'factual' sources.
> (1989:127)

10 Including, for example, the one in which Aitkin (1997) was speaking from: cultural geography; here landscapes and places were now included as versions of these new cultural "texts" that could be read by the scholarly detective.

11 As such, any book is not a straightforward artefact existing in an "object world", but rather is in "a state of . . . constant . . . rema[king]" (Hornsey, 2012:675) and part of an "event world" (following Ong, 1971, and Pocock, 1989), as Hones later acknowledged and developed (2014). To paraphrase Revill, a book is, therefore:

> co-produced by the act or processes of making, the materials which carry and transmit, and the means of receiving, sensing and making sense. [A book is] made within the contingent interplay of each of these realms simultaneously.
>
> (2016:253)

12 Such mutability democratized the critical and popular discussion of literature. The ability to call forth different meanings and thus influence the understanding and definition of texts recognized the role of all readers, be they scholars or the public. In line with feminist currents within the social sciences (see, for example, Haraway, 1988; Bondi, 1990; McDowell, 1999), it emphasized that readers are no longer modern, but humans with politics, emotion, rigour, and imagination which influence their engagement (and we will return to the impact of imagination on literary geographies below). Emplaced within such a current, an interest also grew in reception studies, exploring the type of reading different audiences engaged with (be they academic or lay), informed by what ideas, and the constructive alignment to their aims and ends.

13 This perspective emphasized how our engagements with specific places influence us and through these processes we come to influence them; through our actions, we form the identity of places, and co-constitutively, places become "deeply woven into the fabric of who we are" (Preston, 2003:XVI). The spatial current thus reminded us that, like all "things", we are defined by geography – we are not simply social, or political be(com)ings, but we are ultimately (and perhaps aggregately) spatial be(com)ings too. Just as texts are nothing without context, humans are not (and never have been) abstract, contextless, or modern (after Latour, 1993): we are all spatial beings now.

14 Considering place relationally has therefore challenged scholars to take forward the emphasis on activities and practices, and resist the inertia of considering places as fixed artefacts. As a consequence, places are framed as dynamic and emergent, at every moment, it is possible for a new place to come into being. So, as Brown (2018) suggests, when we consider a specific location, we are no longer looking for one, but rather acknowledging there will always be a plurality. Yet in a similar way to the influence of context on culture, this plurality of place has a further, more complex, dimension. Just as it is acknowledged that due to the influence of the reader there can never just be one (interpretation of a) book, this plurality of place is made more complex by the situatedness of its inhabitants. Considering place relationally has challenged scholars to take forward the emphasis on individuals and their activities, and how their positioning will influence how places are defined. In Jones' words,

> 'thinking space relationally' . . . insist. . .s on [place as] an open-ended, mobile, networked, and *actor-centred* geographic becoming.
>
> (Jones 2009:5, my emphasis)

Such thinking has direct consequence for how we record and register what a place is. As we move away from one fixed or Archimedean position from which to see a location, we accept that there are multiple actor-centred positions we could take; rather than simply "looking for one" (after Brown earlier), we may instead be seeking to capture the infinite (after Solnit, 2010). As Solnit explains, every actor "possesses his or her own [perspectives] of [a] place, a world of amities, amours, transit routes, resources, and perils" (2010:3). Such acknowledgement proliferates the possibilities for literary geographic inquiry and how to grapple with, understand, and communicate these infinite worlds.

15 Despite broader challenges to such "god-like" positioning (see earlier), they seek to offer a view of the world "read literally from above. Insofar as it represents a terrain, that terrain is seen from the heavens" (Idhe, 1990:67).

16 The literary scholar Walford Davies concurs; in his words, the literary text "has always" had the capacity to "function . . . as a map" (2012:12). Indeed the broad category of "cartography" can and should, from his point of view, include not simply "literal physical entities" but echoing Tally's "manner of speaking" should also refer to "verbal choreographies and textual and visual figures" (2012).

17 Deep mapping may seduce the unwary into framing places as durable and essential, suggesting that as long as a scholar compiles enough data over a long enough time frame, one can depict them absolutely and comprehensively (see Cresswell, 2019).

18 As Susan Naramore Maher puts it, deep maps project worlds that "limn a matrix of communal and personal memory, competing histories, and polyvocal narrative" (Naramore Maher, 2014:xiv), as she continues:

> Like a literary survey team, these [deep cartographers] map multiple measures. Inspired by geologists, [they] create cross-sectional narratives of natural history, illuminating the strata in which deep time and human time collide . . . Following cultural geographers, they mark the shifts and migrations, booms and busts, erasures and additions, always keeping an eye open for the palimpsests of former worlds.
>
> (2014:15)

And as Harris confirms:

> A deep map . . . interweaves physical geography and scientific analysis with biography, folklore, narrative, text, memories, emotions, stories, oral histories, and so much more to contribute to a richer, deeper mapping of space and place.
>
> (2015:39)

19 Within this working definition, it is possible, for example, to approach a novel as an isolated artefact if the chosen research questions warranted it, and philosophical position demanded it. Although that work wouldn't share the positioning of the *Literary Atlas* (as one example), they could nevertheless coexist under this definitional banner.

20 With this acknowledgement in mind, it remains vital for all working within the subject to position themselves in relation to what sort of geographical reading they are interested in producing, with what fictions, through what methods, and for what audiences.

4 New literary geographies

An emergent *Literary Atlas*

Introduction

As we have seen, this book argues that literary geographies in their current form emerge at the confluence of many disciplinary approaches. These disciplines have been influenced by a broad sea change away from the modern constitution and towards an ontology of instability, fluidity, and emergence. This sea change has mobilized a range of currents (of "culture", "space", "cartography", and "creativity", see Chapter 3) which have encouraged these disciplines to look beyond their traditional limits and identify creative opportunities to engage with each other.

As a consequence, literary geographies is experiencing what could be understood as perfect storm conditions. A range of approaches are coming to the subject which add vibrancy and potential, opening up new concepts to engage with the possibilities of the written word, the combined spatialities of human life, as well as enabling new cartographic creations to communicate these relations to a range of audiences. This book maintains that literary geographies can harness these unique conditions for innovation, as well as provide an intellectual space for a range of approaches, through adopting an open working definition outlined in Chapter 3. This working definition provides the basis for the *Literary Atlas* project's own approach to literary geographies, which is introduced in this chapter.

A *Literary Atlas* emerges: three irruptive events

In line with Chapter 3's working definition of literary geographies, and its call to be transparent about origins and directions, it is important to identify that the *Literary Atlas* project emerged through three "irruptive events" (after Badiou, 2005, see also Chapter 2).

Event one: reading literature (for cultural geography)

Through my own work in cultural geography, I wished to expand the means through which students could recognize the importance of both culture and place to a sense of human identity. Motivated by the cultural current (see Chapter 3), I had consistently used a range of what I had come to frame as cultural "traces" (see Anderson,

DOI: 10.4324/9781003321248-4

2010, 2015, 2021) to this end (e.g. graffiti, fashion, slogans, adverts, films, songs, music videos, and performances). However, through my own experience of living in a new city and coming to understand it through the poems, short stories, and novels that were set in its streets, I had a growing sense that literature had a power which could beguile, stimulate, and resonate with a sense of (un)belonging in a place, but – more than this – also come to co-constitute how this place was experienced and defined from the perspective of the reader.

At this time, I also developed a project which explored the ways in which literary fictions, written in English and set in Wales, could be "read" through a range of cultural geography theories (see Anderson, 2014). As part of this project, I began to experiment with how we might begin to explore the uncanny means through which reading literature can influence and invent a sense of location. I realized through this process how the worlds created through reading literature could often be partially or completely "out of order" in relation to the modern (after Qvistrom, 2007, see also Lahiri Dutt, 2013), with reading having the power to suspend and subvert conventional definitions of the "real" and "imagined" (see also Chapter 1). As noted in Chapter 1, I experimented with the notion of "plot" to describe this process, and became intrigued as to how it might be used more directly and interrogated more thoroughly in a project beyond the confines of cultural geography.

Event two: writing Britain (?)

As I tentatively explored the possibilities of plot, I was moved to attend an exhibition at the British Library in London in 2012a titled: "Writing Britain: Literature and Landscape". The exhibition sought to display the ways in which literature has been "inspired by spaces and places around the British Isles" (British Library, 2012a). The exhibition included many examples of how the relations between literature and place could be curated for the general public – through films, artefacts, images, and music – and affirmed the possibilities that "creative cartographies" might offer to this end (see Chapter 3). Part of this curation included recordings of writers talking through their understanding of the relations between literature and place. A key insight was provided by Robert McFarlane, who stated:

> almost every significant rock or landmark [in Britain] . . . has stories associated with it, we live in a densely, historically, storiated landscape; how do you find new ways of seeing those, joining those stories up, making songlines as it were, joining place to place, story to story, and image to image?
>
> (British Library, 2012b)

For me, these words compounded the urgency to explore the possibilities of plot – I received the phrases "joining place to place, story to story" as indicating the possibility that reading had the capacity to create an interweave of complexity so "dense" that our "storiated landscape" was not an independent exercise in decoration, but rather a statement of interdependent integrity.[1]

However, despite the inspiration drawn from the "Writing Britain" exhibition, it was also clear that the literature chosen for curation at this event was not representative of all the British Isles. It was notable, for example, that writings in Welsh, or set in Wales,[2] were not included on this platform. For someone who had revelled in these writings for the previous decade, it brought into focus the power of the "canon",[3] and also its gatekeepers. On this occasion, it was clear that when the British Library platformed "Writing (in and of) Britain", it could be taken to mean "Writing (in and of an) England", and through this sleight of geography, a number of intriguing and significant plots had effectively been silenced.

Event three: stakeholders assemble

The third transformative event from which the *Literary Atlas* project emerged occurred as the "Page and Place" project (Anderson, 2014) proceeded. I met with a number of key stakeholders who had committed their professional lives to the study and support of Welsh writing in English, and also shared a passion for the potential to catalogue, document, or otherwise "map" these literatures for posterity and new publics. These included Bronwen Price at Literature Wales,[4] Sarah Morse (based first at Swansea University and then The Learned Society of Wales), and, perhaps most significantly, Kirsti Bohata, from Centre for Research into English Literature and the Language of Wales (CREW), Swansea University. With these stakeholders, I discussed ideas for a project which could be based around Welsh writing in English, and be competitive for Arts and Humanities Research Council funding. Following a successful bid, the *Literary Atlas of Wales and its Borderlands* commenced, assembling Kirsti Bohata as Co-Investigator, Kieron Smith (as an expert in Welsh writing in English, and Research Associate), and Jeffrey Morgan (as a technical specialist in digital mapping, and Research Associate). Scott Orford (from the Welsh Institute of Social and Economic Research and Data, Cardiff University) and Damian Walford Davies (then Head of the School of English, Communication and Philosophy, Cardiff University) became formal mentors to the project, and Literature Wales a project partner.

It is important to note that the *Literary Atlas* team was therefore assembled from beyond one discipline. At first glance, it could be seen to align with an emerging consensus towards a "double-disciplined" literary geography (see Chapter 2). For the *Literary Atlas*, key academics were established in the distinct "towers" of "human geography" (Anderson) on one side and "literary studies" (Bohata) on the other. However in practice, these academic positionings were far more "unstable" (following Tally, 2019, with respective to literary geographies more generally) than at first appeared. Human Geography itself is often practiced as an interdiscipline (see Donaldson et al., 2010, and Chapter 2), and due to my own employment in a School of Geography *and* Planning, and enjoying a sub-disciplinary predisposition to cultural geography, any straightforward reading of my disciplinary identity was complicated. For Bohata, although a scholar of literary studies, she enjoyed specialism in historical fiction (a sub-discipline informed by the disciplines of literature and history), and in particular historical fiction set in Wales; thus her expertise was further complicated by scholarship in, and awareness of, the discipline of Welsh studies. Thus when considering the lead investigators, the *Literary Atlas* project benefitted from a much wider disciplinary skill set than just human geography and literary studies; and when the broader research team was also considered, it was further influenced by and schooled in digital advances in sociology, politics, and the social sciences more generally (Morgan), in Welsh Studies, cultural studies, and documentary journalism (Smith), and informed by the expertise of Literature Wales in place-marketing, publishing, cultural heritage, and literary tourism (Price). This coming together of approaches and expertise thus meant that the *Literary Atlas* team was not confined to a highwire slung between just two disciplinary towers (see Chapter 2), but could adopt an experimental "functional" interdisciplinarity (see Chapter 2) to explore the power of plotted reading to map literary fictions in Wales.

Literary Atlas was thus a project informed by a plurality of disciplinary expertise but led through an interest in literary studies, cultural geographies, and Welsh Studies. It sought to engage with the creative tensions in this confluence of interests when conducting its initial aim, to explore the power of reading literature to influence and invent a sense of location. Expanding the notion of "plot" to enable this process (see Chapter 1), it was motivated to explore to what extent read-realities should continue to be marginalized as esoteric entertainments, or be fundamentally integrated as part of our interdependent ontological experience. To this end, it committed to produce a "book of maps" which sought to experiment with not only how read-realities may be captured, but also how they might be communicated to a range of audiences. This commitment resulted in a specific focus on 12 novels, from which a range of innovative artistic and digital cartographies were created, commissioned, and curated. The project then evaluated these innovations, through working with reading groups, conducting field trips and digital surveys, curating exhibition tours, and engaging the public in writing and mapping workshops. Reflecting both our interdisciplinary expertise, and to draw attention to the silences echoing around the Writing Britain exhibition (earlier), *Literary Atlas* also found its focus through identifying, then broadening, the canon of writing in English.

Identifying and broadening the canon

From their respective (inter)disciplinary backgrounds, the *Literary Atlas* team was strongly aware that the history of Welsh writing offered a wide, diverse, and richly textured set of fictions to focus the project. As such, we were committed to focus on writings related to Wales, but it was nevertheless necessary to make difficult choices over which literatures to include, whilst also being acutely sensitive to the significance of any exclusion. The importance of language was a key factor. Wales as a country has two official languages (Cymraeg and English), and both languages have a strong literary tradition in all forms of writing. Yet English remained the most common language in Wales, the project teams' first language, and the arena of their professional expertise. Although as a collective, we did not wish to explicitly marginalize the vibrant literary history in Cymraeg, we nevertheless acknowledged our lack of "literacy" in this language, and – at a pragmatic level – its relative lack of traction to a wider audience beyond Wales. We therefore committed to focus on Welsh writing in English, but concomitantly flag connections to the Cymraeg tradition wherever it was possible to do so (through, for example, initial scoping choices which favoured writers that worked in both languages, and choosing literary fictions that historically forged strong linkages to Cymraeg literature[5]).

The issue of language confirmed to the *Literary Atlas* team that any project could not – and should not – claim to be definitive or comprehensive in its nature: ours, therefore, was not *"The" Literary Atlas of Wales*, but simply *"A" Literary Atlas*. Indeed, as literature not only comes in many languages but also many forms, we were also faced with identifying which forms would be included in our *Atlas*. Welsh writers may be internationally well-known for their plays and poetry (for example), yet our interests consistently returned to the long-form story. With these professional interests in mind, the novel became our focus, with the options remaining open for future projects to extend into Cymraeg at one level, and short stories, plays, or poems (for example) on another.

The final macroscoping criterion that was considered at this early stage was the geographical focus of the novel. We were excited by the prospect of centring exclusively on novels that were predominantly based in Wales, with Welsh places or regions central to the narrative. This focus meant that novels could be included even if their author was not Welsh (by birth or residency), or the book was published through a non-Welsh publisher; but similarly novels could not be considered even when written by a Welsh author if its location did not correspond in some form to one in Wales.

From these initial criteria, we sought to identify how many published novels would fit our criteria. To do so, we negotiated access to key archives that catalogued Welsh writing in English. These included the Celtic Collections in Cardiff University, the bibliographic catalogues of Welsh Writing in Swansea University, and the comprehensive archives of the National Library of Wales (Aberystwyth). Although these catalogues did not necessarily strictly align with the project's criteria,[6] when aggregated they nevertheless offered an insightful context of the possible novels which could be chosen for our *Literary Atlas*. In order to help the selection process, an aggregated catalogue was created for each possible entry. Every novel was noted in bibliographic terms (with its author, title, and date noted, as recorded by the respective library sources). Although not a direct motivation for our interest, with this data it would have been possible, for example, to analyze the gender of authors, the historical development of the novel as a form, alongside the development of publishing as an industry in Wales (see www.literaryatlas.wales/en/library/). In total, we identified 573 novels, published from 1800 to 2018.

This baseline was useful to the *Literary Atlas* team as from it we could identify the key geographical distribution where these novels "took place" (see Anderson, 2021, but also, for example, Reuschel and Hurni, 2011:218). If we could identify the locational distribution of our catalogue, this could become a useful "provocation" for our decision-making (Piatti, 2016:91), providing us with a criterion against which to choose which novels we would focus on in the project as it progressed (see below). To do so, we initially engaged in a "geoparsing" exercise. As Gregory and Donaldson identify, geoparsing is often the "first step" in locating literature. As they put it:

> Geoparsing involves identifying and extracting place-names from the corpus under analysis and assigning each place-name to a co-ordinate location.
>
> (2016:69)

To this end, the *Literary Atlas* team manually located each novel in its respective library, and from studying its key information (for example, its synopsis blurb, publisher commentary, or targeted reading of its text), identified and sense-checked a "candidate" (after Weller, in McLaughlin, 2023) for the principal location in which the novel took place. This candidate may have been "slippery" in nature (after Bushell, 2012), in other words, the author may have used a pseudonym for a location and, therefore, a degree of informed judgement was required to determine which location was a likely corresponding geography. At this stage, however, it was possible to work with a degree of approximation (or what Bushell terms a "nebulous fluidity" (2012:272)) as our aim was not for geographical accuracy as an intrinsic goal, but rather a general sense of the areas of Wales covered by the catalogue in aggregate.

This exercise in geoparsing was not therefore a strict act of close, spatial, or indeed plotted "reading". It rather echoed the Morettian tradition of distant reading (see Chapter 3): we "reduced" the text to its dominant location, "abstracted" this from the narrative flow, and then sought to "construct a new, artificial object" from this process (2013:53). Indeed, many projects undertake similar analyses using automated algorithms and, at this level of simplification, this process requires little "detailed textual engagement" (after Gregory and Donaldson, 2016:270), rather a "distanced" overview. As a consequence, any distant reading does not account for a place's dynamic "locale" (after Agnew and Duncan, 1989), or its cultural and literary significance to the plot (although see below), nor could it offer – due to any potential omissions in the source catalogues, or due to inaccuracies in interpreting "slippery" locations – an incontestable cartography of, in this case, all English-language fiction in Wales.

For the *Literary Atlas* project, therefore, geoparsing's function was to create a set of data which listed key candidate locations for English language novels taking place in Wales, as listed on the catalogues of Cardiff University, Swansea University, and the National Library of Wales. The *Literary Atlas* team then had to decide how to engage with this data. We could, for example, sift and filter the catalogue by place, to ascertain the quantity of novels in a certain location, and identify patterns in relation to the date published or author. It is important to note that, in the broadest sense, this would be one variety of analysis of literary geographies.[7] However, the *Literary Atlas* project was more interested in discovering what "emergent qualities" (Moretti, 2005) may be revealed through different visualizations of this catalogue. As Gregory and Donaldson note, visualizations are often conducted following a geoparsing process, and such moves commonly involve graphically charting co-ordinate locations on some form of map, as they put it:

> Once the corpus has been geoparsed . . . , the georeferenced place-names it contains can be extracted . . . and imported into a GIS application [for example] where it can be displayed and analysed.
>
> (Gregory and Donaldson, 2016:69)

This process of connecting different distant representations together (i.e. a catalogic corpus and conventional geographical map) can be understood as a process of "translating one form of literary cartography into another" (Anderson, 2016b:162). This process is undertaken in order to invite new insights into the data through, literally and figuratively, seeing this data in a new way. In a similar vein to visual poetry (which is written specifically on the page so its form can add meaning), or indeed in line with Benjamin's "parataxis", our catalogic corpus and conventional cartography could be "simply juxtaposed" (after Cresswell, 2019:6), in the hope that this new display may "provide flashes of recognition" (Cresswell, 2019:9).

In order to enable the catalogic corpus to be connected to a conventional cartography, Google Maps API was used to convert the data into an interactive map. Each digitized geotag was pinned to the relevant geographical co-ordinates, with the entry details made available to the user via a clickable icon which generated a pop-up data window. This distant map could present the project with a broad-brush impression of the geographical distribution of fiction set in Wales (and you can find this map here: www.literaryatlas.wales/en/library/).

Mapping this geoparsed catalogue presented the project with the opportunity to see the compiled data differently. Through offering a specific perspective – a

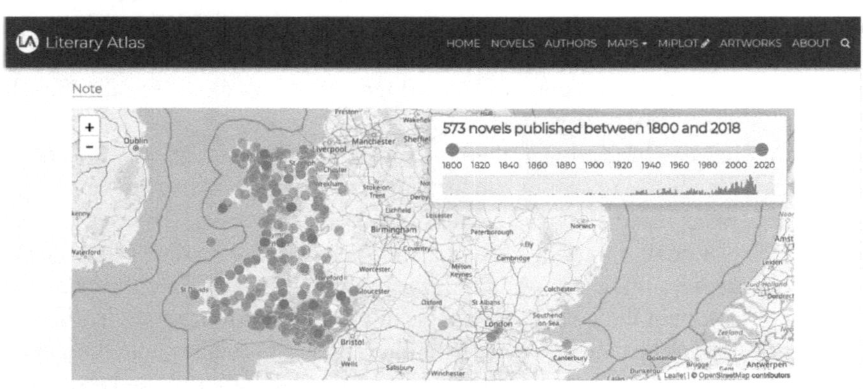

Figure 4.1 Bibliographic catalogue mapped at the national scale.

Source: Key. Light geoparsed dots indicate a novel's associated plot, with dark dots indicating plot for shortlisted novels.

reduced and abstracted "view from above" (after Rosenbaum, 2001:56), this new view was not a mapping of the novels themselves, but rather a constructed "artificial" object (after Moretti, 2005) of candidate location, seen from an almost "heavenly" position outside both the catalogue (Rosenbaum, 2001:56), but also the lived geography of Wales. It was evident, however, that this translation of literary catalogue into literary distant map gave a different "feel" to the data. Through this translation of literary cartography, we could identify that the geoparsed locations of books were diverse. From our broader knowledge of the nation of Wales, it could be seen that there was some clustering around key population centres, and often around key transport routes (principally road), most obvious in the valleys of South Wales. Similarly, there was a paucity of locations in the gaps between these routes and centre (e.g. between Welshpool and Commins Coch). From this platform, it would be possible to add in layers of new context to explore, for example, the strength of correlation with respect to key historical periods (for example, the Industrial Revolution), or related to population densities or transport routes.

Thus the conversion of 571 candidate locations into abstracted numbers, then transposing them onto an abstracted digital grid of Wales, offered some degree of insight into the bibliographic catalogue of Welsh writing in English. This distanced perspective allowed the intrinsic geography of the catalogue to visibly emerge, the representative function of the map, along with its interactivity (initiated through touchscreen or keypad), enabled the user to "find themselves" with changes in scale and orientation. From this new view it was possible to identify places of interest or familiarity, engage with unusual names, or dates of significance; the *Literary Atlas* team could ponder authors known and unknown, connect eras and titles and authors in particular places in odd marriages of geographical coincidence. In this way, reading the distant map attuned ourselves as users to the "palimpsestic" or "stratigraphic" (after Mitchell, 2011:73) processes emergent within what appears at first sight to be a superficial, thin, and distant artefact. However, it was not simply layers of catalogic knowledge that were brought into conversation with this new cartographic representation. Interactive "navigations" also mobilized the users' existing knowledge. Connections were prompted by the project team's lived or trained understanding of specific regions of the country (e.g. where key cities were located, or where the main transport routes were), whilst also flagging gaps in this literacy (we asked, for example, "what's *that* book?" or "where is *that* place?"). Thus although this may have been a distant map, it did not provoke in us a distant reading: our navigations, or readings, began to resemble plotted readings (see Chapter 1), as "matri[ces] of communal and personal" (Naramore Maher, 2014:XIV) knowledge converged with the cartographic representation, and different onto-epistemological knowledge fragments merged into one, multidimensional mosaic. We became fully sensitized to how our existing knowledge of distant maps and how they are composed (with legends, keys, scales, and projections), and the freedom they offered our "reading" (not necessarily required to be left to right, top to bottom, like a modern book, for example). We were reminded of our "deep" awareness of conventional (physical and political) cartographies, and our ability to broadly recognize coastlines and borders whilst orienting our geographical knowledge to these representational lines and colours. This knowledge provided the "foundations" for the positioning of geoparsed locations presented on the map, giving us prompts to identify and interpret clustering or other patterns, and how these may suggest correspondence with other processes (e.g. historical or industrial development).

Thus, our engagement with this map, because of this coincidence of abstract data, cartographic representation, and personal connection, was far from being cold and thin, or even simply a generative process for rational "thought" (after Moretti, 1998). Reading the distant map became an experience that was centripetal, drawing us into a world where a range of knowledge fragments (words, numbers, grids, tacit knowledge, academic insight, and flights of fancy) coexisted in, and co-composed a non-hierarchical ontological plane. In Harmon's words, the creation of this simple distant map "intrigued" us (2003:10); it became:

> a vehicle for the imagination, fuelled up and ready to go. We look[ed] at these maps, and our minds kn[e]w just what to do: take the information and extrapolate from it a place where [we] can leap, play, [and] gambol.
>
> (2003:10–11)

Thus, despite any concern that a thin, distant map may produce a thin, abstract(ed) reading, for the *Literary Atlas* team, it produced an "uncanny" experience (see also Walford Davies, n.d.). Our background literacy of the texts, our knowledge of their bibliographic information, their geoparsed locations, and now the visualizations of their candidate locations on a national map of Wales produced "a powerful, semi-enchanted sense of co-presence, of being in [more than] two places at once" (following Moran, 2022:115). However, we were fully aware that, due to their relative "thin" nature, the provocative capacity of distant maps relied in no small measure on the richness of the "matrix" (Naramore Maher, 2014) that the reader could bring to it. Given our disciplinary training and personal interests, we could assume that we might find this distant map more "intriguing" than others. Nevertheless, this process confirmed to us that, if future maps emerging from the project could provide "deeper" materials to intrigue their audiences, it was possible that many users could similarly experience the capacity of maps to become "storytelling stimulators" (Caquard, 2011:6). The effectiveness of a map could be judged by new criteria: how well could a map generate "relational knowledge" (Dickens and Dickens, 2022:23) and a sense of co-ingredience between users, texts, and places (see Chapters 5 and 6)?

From catalogues to canons

As stated earlier, a key motivation for mapping the library catalogues was to help inform decision-making over which novels could be included in further mapping exercises within the *Literary Atlas*. As we approached this decision, we reflected on the *New Welsh Review*'s own choices for the "Best Welsh Novels of all Time" (a process undertaken in 2014). Twenty-five novels made their provocative and inventive long list,[8] as they explained:

> While classics like Raymond Williams' *Border Country* and Emyr Humphreys' *A Toy Epic* have earned their rightful places on this list, you'll also find some more unexpected entries; *Downriver* by Iain Sinclair and Diana Wynne Jones' *Howl's Moving Castle* to name but a few.
>
> (2021)

To some extent, *New Welsh Review* was a (popu)list, appealing to a range of audiences (including Dahl and Dunthorne, perhaps for younger or newer readers), demonstrating a range of genres and literary forms (novellas and novels), and mixing the contemporary with the "classic".[9] This list included a mix of languages, with some written in Cymraeg, some in English, and some available in translation (indeed *New Welsh Review*'s "best novel", Caradog Prichard's *One Moonlit Night (Un Nos Una Leaud)* was one such (1961)). What was insightful from this list, with direct regard to *Literary Atlas*, was that in line with the working definition of literary geographies offered in Chapter 3, it was possible to critically engage with the "canon" of Welsh writing in English to inform and perhaps reform it, in the contemporary era.

As stated earlier, the *Literary Atlas* team was strongly aware that the rich history of Welsh writing in English was often overlooked by an English canon, so often defined by a "familiar roll call of names" that had become "dyed into the fabric of British" literature (Boakye, 2023). We were also sensitive to the possibility that a similar process may be occurring within Welsh literature.[10] It was decided, therefore, that in order to draw attention to new words and voices alongside established works, our final list for the *Literary Atlas* should include both "classic" Welsh fiction and contemporary writers. We wanted, ideally, to combine a mix of genres, writers, and geographies (informed by our catalogic cartography). From these broad criteria, the *Literary Atlas* team consulted with a range of academic and literary experts in Wales, as well as asking the public to offer their own choices for selection through social media. From this, a long list of novels was drawn up from the Welsh Collections' databases.[11] From our long list, a final short list of 12 was drawn up, which included (in alphabetical order by author surname):

1. Trezza Azzopardi, *The Hiding Place* (2000)
2. Alys Conran, *Pigeon* (2016)
3. Fflur Dafydd, *Twenty Thousand Saints* (2008)
4. Amy Dillwyn, *The Rebecca Rioter* (1880)
5. Menna Gallie, *Strike for a Kingdom* (1959)
6. Alan Garner, *The Owl Service* (1967)
7. Niall Griffiths, *Sheepshagger* (2001)
8. Tristan Hughes, *Revenant* (2008)
9. Lloyd Jones, *Mr Vogel* (2004)
10. Christopher Meredith, *Shifts* (1988)

11. Malcolm Pryce, *Aberystwyth Mon Amour* (2001)
12. Raymond Williams, *Border Country* (1960).

As we can see, 8 of these novels were published after 1980, with 7 of those published in the twenty-first century, 1 from the nineteenth century, and 3 from the 1950s–60s. There are 5 authors who identify as female, with a range of provenances and backgrounds (Trezza Azzopardi has a part-Maltese heritage, Niall Griffiths and Alan Garner were born in England, and Malcolm Pryce grew up in Aberystwyth, but has his lived adult life outside Wales). These novels exemplify a mix of genres, and as we will see, a range of regional locations in Wales (www.literaryatlas.wales/en/ but also Chapter 5). The remainder of this chapter will introduce these selected novels, before Chapter 5 begins an extended exploration of the novels, and their plots, in depth.

Introducing the novels

Trezza Azzopardi, *The Hiding Place* (2000)

The Hiding Place (2000) was Trezza Azzopardi's first published novel. It tells the story of the Gaucis, a half-Maltese, half-Welsh family living in Cardiff. The story is told through the first-person narrative voice of Dolores, the youngest of 6 daughters of Frankie Gauci, a Maltese immigrant, and his wife Mary. It is the late 1990s and Dolores has returned to Cardiff to attend the funeral of her mother and reconnect with her sisters. Her return to the locations of her childhood evokes vivid, traumatic memories and perspectives on her upbringing in the 1960s. The novel is located in Butetown – an area near the docklands area of Cardiff known colloquially as "Tiger Bay". Tiger Bay famously supported a rare kind of multicultural community due to its close proximity to Cardiff Docks, a busy port during the height of the Welsh coal industry in the late nineteenth and early twentieth centuries. The shipping industry brought in migrants from around the world to work and live in Cardiff, and the result was a richly multicultural community populated by people of many ethnicities, languages, and religions. For Dolores, Tiger Bay is a place of gaming rooms and cafes, of crumbling houses and burning secrets. *The Hiding Place* was shortlisted for the Booker Prize in 2000.

Alys Conran, Pigeon *(2016)*

Pigeon is a novel set in an unnamed slate-quarrying town in North Wales. The town lies on the periphery of the Snowdonia National Park, and its cultural landscape is dominated by the juxtaposition of the rugged beauty of this topographical surface and the quarries that extract the geological riches of slate beneath. Although the town is not named in the text, Alys Conran acknowledges its location is based on Bethesda. This town and its region is the area in which Conran grew up, and following a period abroad, has returned. This story of growing up, exile, and homecoming finds a parallel in the themes dominating the novel, and in particular, the title character. Pigeon, and his friend Iola, are two young people dispossessed from the constraining familial and cultural structures of their town, and how the power of language may bring them home. *Pigeon* was Wales Book of the Year 2017.

Fflur Dafydd, **Twenty Thousand Saints** *(2008)*

Viv and Delyth have taken their young sons to live on Ynys Enlli/Bardsey Island following the failed 1979 campaign to achieve devolution for Wales. Now it is 2007, and much has changed. Politically, Wales voted for devolution in 1997, whilst personally, Delyth has disappeared and Viv's son, Iestyn, initially convicted for Delyth's presumed murder, has now been acquitted. On the island, Deian, Delyth's son, is undertaking an archaeological survey, whilst a writer in residence, Mererid, along with a film crew are making a documentary. In the hothouse conditions of the isolated island, the real story of Delyth's disappearance is revealed and island inhabitants – old and new – are forced to reevaluate their relationships, beliefs, and ambitions. *Twenty Thousand Saints* is the author's own adaptation of her Cymraeg language novel *Atyniad. Atyniad* won the Prose Medal at the 2006 National Eisteddfod.

Amy Dillwyn, **The Rebecca Rioter** *(1880)*

Amy Dillwyn's *The Rebecca Rioter* is a radical retelling of the Rebecca Riots, a series of popular uprisings which swept through West Wales and into Glamorganshire in 1843. In these riots, farmers and labourers dressed in women's clothes and destroyed toll gates which taxed their routes from farm to market. The novel, written in 1880, is based on these historical events and in part on the account penned by the author's own father, Lewis Llewelyn Dillwyn, a magistrate, who had been amongst 11 men who were ambushed by rioters at Pontardulais in 1843. While the main plot revolves around the riots, the tension between classes is humanized through the developing relationship between a landowner's daughter, Miss Gwenllian, and a radicalized labourer, Evan. The novel is subtitled "A Story of Killay Life" and the principal setting is Upper Killay, near Swansea, and the commons, moors, and woods to the West.

Menna Gallie, **Strike for a Kingdom** *(1959)*

Strike for a Kingdom is set in a Welsh mining village during the "Angry Summer" of the 1926 Miners' Strike. In May of that year, the Trades Union Congress called a General Strike in support of British miners faced with wage reductions and longer working hours. The General Strike lasted 9 days, but miners stayed out on strike for months afterwards. The novel is located in the village of "Cilhendre", a semi-fictional location based on Gallie's home village of Ystradgynlais in the Swansea Valley, which was, at the time of the strike, home to numerous coal mines and heavily dependent on this industry. *Strike for a Kingdom* takes the form of detective novel, in which the local Pit Manager is found murdered. This "whodunit" plot, focusing on a small location and bringing in a cast of characters from across the social classes, enables Gallie to provide a snapshot of a Welsh mining village during this fraught moment in Welsh social history.

Alan Garner, **The Owl Service** *(1967)*

Alan Garner's *The Owl Service* is a novel which reinterprets the Fourth Branch of the Mabinogi, a set of ancient and multi-layered Welsh myths.[12] *The Owl Service* retells a section of this complex story centred on Blodeuwedd – a girl made from meadowsweet flowers. Blodeuwedd's destiny was to be part of a destructive love triangle, and be eventually transformed into an owl. Garner's novel centres around three teenagers who appear to be caught in a Blodeuwedd-style love triangle. Alison and her stepbrother Roger visit Wales from England with their family on holiday; they meet Gwyn, a local youth, whose mother keeps the holiday home in which Alison and Roger are staying. Together they discover a set of plates with an owl insignia on them: could the discovery of this "owl service" ensnare the valley once again within an ancient curse? Garner chose to locate his retelling of Blodeuwedd's story approximately 20 miles to the south-east of Tomen y Mur in Llanymawddwy, a settlement surrounded by the Aran mountain range. *The Owl Service* was the winner of both the Guardian Award and the Carnegie Medal.

Niall Griffiths, **Sheepshagger** *(2001)*

In this provocatively titled novel, Niall Griffiths explores themes, characters, and locations introduced in his first novel Grits (2000). His characters inhabit the underbelly of "Generation X" culture, living on benefits, their lives defined by alcohol, drugs, and casual violence. Into this mix, Griffiths introduces Ianto. Unlike his characters who originate from industrial South Wales or Northern England, Ianto is indigenous to the rural, wild hinterland of Aberystwyth. Ianto was born and brought up in the remote mountains by his Mam-gu (Grandmother). From an early age, Ianto displays a strong affiliation to the brutal beauty of the area and its non-human species, and – just as strongly – against any people or practice that threaten them. This loyalty and connection is intensified through a number of violent childhood experiences, and becomes unleashed when his family home is repossessed.

Tristan Hughes, **Revenant** *(2008)*

After 10 years apart, 3 friends return to the small Welsh town in which they grew up to revisit a site that binds them all. On the way, they are haunted by memories and apparitions, and their visitations stir powerful, and sometimes disturbing, emotions. As with Hughes' earlier works, *The Tower* (2004) and *Send My Cold Bones Home* (2006), Revenant is set on the island of Ynys Môn (Anglesey), off the coast of North Wales. Though the town is not named, the novel's descriptions of local landmarks suggest that much of the novel is set in and around a strip of coast on the eastern edge of Ynys Môn, incorporating the seaside town of Beaumaris, and the villages of Llangoed and Penmon.

Lloyd Jones,* Mr Vogel *(2004)

Mr Vogel is the labyrinthine story of a small country and a lame man's quest for a lost bear. The novel generates its own mythology as it circles in fantastical and comic fashion around the simple story of a children's ward in Shropshire in the late 1950s, and a pub in Llanfairfechan, North Wales. It weaves together stories of travellers and mystics, drunks and detectives. Characters move in and out of wormholes in time, a man builds a stairway at the bottom of his garden before climbing it and disappearing. Mr Vogel takes its characters on a circumnavigation of Anglesey and then a walk around the whole circumference of Wales. The different locations become historical palimpsests through which the author stitches together both the disguised story of childhood trauma and a comic celebration of the culture of Wales.

Christopher Meredith,* Shifts *(1988)

Shifts is set during the long closure of one of the major industrial employers in South Wales: the Ebbw Vale Steelworks. It follows 4 main characters and their attempts to make sense of their lives in a rapidly de-industrializing community. Keith is looking for meaning in history, looking back to the early years of industry in the valley. Judith is focused on the present, though she is hardly satisfied with life on the housing estate or the role of housewife to an unexciting husband. Jack is escaping his recent past in England, but is reticent about coming home. O is an observer; ritual in his rigid adherence to habit, he is also a resentful outsider. *Shifts* is an examination of the relationship between our inner lives, the people around us, and the forces of history.

Malcolm Pryce,* Aberystwyth Mon Amour *(2001)

There are 6 books in Malcolm Pryce's "*Aberystwyth*" series, each recounting the travails of private detective Louie Knight and his sidekick Calamity Jane. In this first book, Louie and Calamity take on the case of Myfanwy Montez, the celebrated singer from the infamous nightclub The Moulin. Myfanwy's cousin Evans the Boot has gone missing, and it seems like a cabal known as the Druids might be behind his disappearance. As its title suggests, Pryce sets his detective series in the Welsh seaside town of Aberystwyth. His interpretation of this location mixes very British notions of seaside holidays – lost innocence, saucy postcards, and donkey rides – with more universally known characters from hard-boiled crime thrillers. This juxtaposition combines to create a blackly comic mystery novel.

Raymond Williams,* Border Country *(1960)

Border Country (1960) was literary scholar Raymond Williams' first published novel. This semi-autobiographical work tells the story of the Price family and their experiences living in a small village in the Welsh border country. As well as being a sensitive, lyrical portrayal of interpersonal relationships – wives and husbands, fathers and sons, friends and colleagues – *Border Country* is a story about a village, and the powerful social and historical forces that impact upon places and the communities who inhabit them. The novel is located in the village of "Glynmawr" in the Welsh border country. Though fictionalized, Glynmawr is closely based on Williams' real home village of Pandy.

Conclusion: making the distant deep

In this chapter, we have charted how the *Literary Atlas* emerged from a concerted attempt to champion not only Welsh writing in English but also the power of reading fiction to mobilize connections to places, and ultimately understandings of the inevitably spatial human condition (see working definition of literary geographies, Chapter 3). Chapter 4 has documented the subsequent journey from catalogic foundations to selected short list through making visible the geographical context which co-defines each novel, and the juxtaposition of this data to a digital map. This process converted catalogic data to (carto)graphic representation, and enabled the *Literary Atlas* team to see Welsh novels written in English in a new way. As Owen recognizes, "maps allow patterns to emerge and become real" (2002:154). Chapter 4 confirms that Owen's choice of verb (i.e. "emerge") is significant; literary maps do not explain relations or nascent patterns, but rather invite the user to identify and reflect upon them; in this way, literary maps become invitations to the user to actively interpret and form their own readings of their data. Maps therefore encourage the user to bring their own tacit and trained knowledge to these "thin" objects, and transform their juxtapositions into inferences and suggestions, the graphic representations encourage curiosity and compel associations to be made with the data set. In short, literary maps provide the basis from which the user can begin to put their own geographical and cultural context back to literary geographies represented. In this way, (even) distant maps can provoke deep engagement, which when explored, enables the geographical human to call forth questions and queries, inviting possibilities and potential "patterns to *emerge*" (after Owen, 2002, emphasis added).

Even at this early stage, therefore, the power of (plotted) engagement between reader and literary geography is suggested. To attend to the second half of Owen's statement (earlier), maps do not only "allow patterns to emerge [but also to] *become real*" (2002:154, emphasis added, see also Chapter 13). By harnessing the power of literary maps to graphically represent bibliographic data and geoparsed locations, a suggestion is offered to the user that these books are rooted in these locations, and the mere suggestion is enough to stimulate a series of questions in the user (for example, is this juxtaposition plausible? What might this mean (for the novel, for the place, and for my relations to them)?). In this way, digital distant mapping has the potential to capture the users' imagination and offers new points of entry into fiction, mobilizing the matrix of personal geographies which co-define the user and assembling it with that of the map. In this way, as Gregory and Cooper suggest, "technologies can be used to manage, manipulate, [and] visualize . . . the geographies embedded within texts" (2013:267) and their users. Although the map may be produced through a modern atomizing process of "abstract, reduce, extract", the map itself promotes a spatiality which stimulates deeper engagement. The power of re-mediating literature in this way confirms the potential for literary maps to offer "more than the sum of their parts: they will possess 'emerging' qualities, which were not visible" in the original (Moretti, 2005:53). With this insight, the question was inevitably posed: how could *Literary Atlas* employ plotted reading to harness this cartographic potential and generate new stimulating insights about literary geographies? In Chapter 5, we explore this question by first positioning the *Literary Atlas* in relation to other large-scale literary mapping projects.

Notes

1 MacFarlane's reference to "storylines", evoking indigenous peoples of Australasia and their amodern ontological relations between "stories" and "places", also created resonance with the development of "plot" into "plotlines" (see Anderson, 2014, 2015, and also Chapter 6).

2 Although the Welsh novelist Owen Sheers was included in the "talking head" videos commissioned to introduce and narrate the exhibition.

3 This absence of Welsh writing in English in the national canon was also emphasized when discussing the key GCSE and A-level curricula, even in Wales. The lack of Welsh writing in English in the novel categories was notable by its prolonged absence (see also Chapter 13).

4 *Literature Wales* is an organisation funded by the Arts Council for Wales to support the skills of writers in Wales, develop communities to participate in literature, and "celebrat[e] Wales' literary culture" see Literature Wales, 2024.

5 We also chose not to engage in the complexities of translation between languages – a sometimes magical act that has the capacity to lose or find new meanings from one language to another.

6 These catalogues, for example, often included some novels whose primary location was outside Wales, as the maps available online indicate.

7 As we have seen in Chapter 3, a novel can be understood as one form of creative cartography, and although unusual, the catalogue of candidate locations could also perhaps be seen as a chart, or map, of literary places.

8 These novels were *Ash on a Young Man's Sleeve* by Dannie Abse (1954); *The Old Devils* by Kingsley Amis (1986); *The Hiding Place* by Trezza Azzopardi (2000); *So Long, Hector Bebb* by Ron Berry (1970); *The Genre of Silence* by Duncan Bush (1995); *On the Black Hill* by Bruce Chatwin (1982); *Fantastic Mr Fox* by Roald Dahl (1970); *Awakening* by Stevie Davies (2013); The Withered Root by Rhys Davies (1928); *Submarine* by Joe Dunthorne (2008); *A Toy Epic* by Emyr Humphreys (1958); *The Valley, The City, The Village* by Glyn Jones (1956); *Cwmardy* and *We Live* by Lewis Jones (1937); *The Great God Pan* by Arthur Machen (1894); *Shifts* by Christopher Meredith (1988); *The Life of Rebecca Jones* by Angharad Price (2002); and *Un Nos Ola Leuad* by Caradog Prichard (1961); *Chasm City* by Alastair Reynolds (2001); *Gold* by Dan Rhodes (2007); *Twyll Heno* by Kate Roberts (1962); *Downriver* by Iain Sinclair (2004); *In and Out of the Goldfish Bowl* by Rachel Trezise (2000); *The Little Stranger* by Sarah Waters (2009); *Border Country* by Raymond Williams (1960); *Howl's Moving Castle* by Diana Wynne Jones (1986).

9 We can see from our bibliographic catalogue earlier that New Welsh Review's choices generally reflect the growth of the novel over the last 40 years (14 of 25 chosen were published since 1982) and the broad socio-demographic of published authors (with 7 of the chosen novels, for example, written by women).

10 With Finch's (1987) parody of RS Thomas' "A Welsh Landscape" (see Evans, 2013), whispering in our ears, "To live in Wales/. . . is to be mown down by the same words at least six times per week" (see also Anderson, 2014:49).

11 Author long list included: Tom Bullough, Nikiti Lalwani, Gwyneth Lewis, Owen Sheers, Sean Burke, John Williams, Tom Anderson, Tessa Hadley, Niall Griffiths, Malcolm Pryce, Richard Collins, Grahame Davies, Amy Dillwyn, Allen Raine, Hilda Vaughan, Iain Sinclair, Gwyn Thomas, Catrin Dafydd, Raymond Williams, Ron Berry, Lloyd Jones, Sian James, Cynan Jones, Fflur Dafydd, Rachel Tresize, Tristan Hughes, Emyr Humphreys, Joe Dunthorne, Christopher Meredith, Stevie Davies, Erica Wooff, Dan Rhodes, Catherine Merriman, Lewis Jones, Menna Gallie, Bruce Chatwin, Glyn Jones, Dannie Abse, Rhys Davies, Trezza Azzopardi, Arthur Machen, Duncan Bush, Richard Llewellyn, Alexander Cordell, Alan Garner, and A J Cronin.

12 Careful consideration went into selecting The Owl Service for the *Literary Atlas*. Due to the important cultural heritage and value of the Mabinogi stories – from the perspective of the *Literary Atlas* team, these myths are as important to Welsh culture as Homer is to early Greek or Chaucer to English storytelling – it seemed an omission too far to not include them in some shape or form. However, the origins of the stories are difficult to pin down; they were first told (rather than written) in the oral tradition, and later transcribed into the Middle Welsh language in the twelfth and thirteenth centuries. In the late eighteenth and early nineteenth centuries, the Mabinogi were translated and compiled into English. From then on, they have proved to be fertile terrain for a range of reworkings and retellings. As this *Literary Atlas* opted to not include first language Welsh or translations in the source material, we looked to these modern reworkings in order to "nod" towards the importance of the original mythologies, and included a novel that has sustained appeal to a range of audiences. In many public engagements early in the project, *The Owl Service* was continually mentioned as a book the project should consider including.

5 Locating the *Literary Atlas*

Developing plot-points

As we have seen in the previous chapter, the conversion of one form of literary cartography (for example, a book) into another (e.g. an abstracted bibliographic catalogue, and then a "distant" map) is a process that can be used to: "manage, manipulate, [and] visualize . . . the geographies embedded" not simply "within texts" (following Gregory and Cooper, 2013:267) but also their users. Each step in this process sets in motion a juxtaposition of knowledge fragments that, when read in specific ways, generates a range of "emergent qualities" (after Moretti, 2005) which produce new insights about the relations between reading, places, and literature, and by extension, our broader connections to the world.

As Chapter 4 has outlined, an initial exploration of this process generated a shortlist of novels for the *Literary Atlas of Wales and its Borderlands* project. From this basis, the *Literary Atlas* team began to extend the exploration of the relations between plotted reading and literary mapping by engaging with these novels in depth. As this process was initiated, like all cartographers, the team faced a set of key "decisions" (after Robinson, 1952) about how to identify relevant extractions from the selected novels, and then, how to map them. In line with the cartographic current (see Chapter 3), it was clear that the decisions that we made would form the world that we mapped. This chapter continues by documenting the discursive process through which the *Literary Atlas* team made these decisions, then detailing the consequences these decisions had for how our new "read-realities" were mapped as a consequence. The chapter begins by outlining how the *Literary Atlas of Wales* team came to position itself as philosophically and practically different from the already-established *Literary Atlas of Europe*, and attempted to realize the opportunities that were afforded by moving away from a distanced and disembodied classification of texts, and towards more plotted and extended identifications of "plot-points". The chapter continues by outlining how the identification of plot-points and their broader co-texts could be used to produce a first set of maps of selected novels, which would form the basis for plotline development that is introduced and discussed in Chapter 6.

DOI: 10.4324/9781003321248-5

Positioning the *Literary Atlas*

To begin developing the precise nature of the *Literary Atlas* approach, the team initially turned to Piatti and Hurni's *A Literary Atlas of Europe* (see literaturatlas.eu, also Piatti et al., 2008; Piatti and Hurni, 2009; Reuschel et al., 2009, 2013) to not simply learn from, but also locate ourselves in relation to, this key work. At the outset, this project appeared notably different from our own. The *European Atlas* demonstrated the conventional "proclivity" (following Braun, 2004, see also Chapter 1) to adopt modern categorizations, "purifying the world" of literary geographies into "real" geographies on one hand (which were also referred to by the *European Atlas* as "geospace" (see, for example, Piatti and Hurni, 2011:218)), and "fictional geographies" on the other (referred to as "fictional space" or "textual space" (e.g. Reuschel et al., 2013:138)). In line with the modern constitution, such separation implied associated privileging: whilst the "real" was fixed, measurable, and predictable (and was therefore aligned itself with onto-epistemological "goods"), the "imagined" was not (being transient, nebulous, and subjective, and therefore onto-epistemologically "other"). Such categorizations were augmented with equivalent sub-categories, including, for example, "settings" and "projected places" (Reuschel et al., 2013). The terminology of these notions did little to contest the (modern) notion that place and drama were only passively aligned (rather than, for example, definitively co-ingredient in their nature, see Chapter 3).

Despite acknowledgement of the philosophical differences between the *European Atlas* and the *Literary Atlas of Wales*, we nevertheless reflected on the utility that the former's key sub-categories may have for our map-making decisions. According to Reuschel and Hurni, the "categorical relations between fiction and place" include a number of key types of "setting", including, for example, "imported", "transformed", and "invented" locations (2011:296). "Imported settings" are those which "correspond . . . with real space by realistically portraying it" (2011). This category, we reflected, may be useful to describe, for example, Fflur Dafydd's portrayal of Ynys Enlli/Bardsey Island in her novel *Twenty Thousand Saints* (2008). Yet, despite its apparent utility, the term nevertheless troubled us. In our reading, the term "imported" referred directly to the practice of bringing components into a place from "abroad" or other perhaps otherwise alien or "strange" (Simmel, 1950) territory. With this implication in mind, it echoed our concerns about the potential inappropriateness of "fit" between categorizations (in this case, the modern philosophical standpoint itself) being "imported" into the experience of reading a novel. Other terms were also troubling. For Reuschel and Hurni, a "transformed" location was a "setting transformed through remodeling, renaming, relocation or synthesis of several places" (2011:296). For the *Literary Atlas of Wales*, a "transformed" location may arguably apply to Menna Gallie's Cilhendre in *Strike for a Kingdom* (a transformation of the "real" Ystradgynlais in Swansea valleys, South Wales) (1959), and to some extent Trezza Azzopardi's Butetown in *The Hiding Place* (in which she relocated streets from nearby Adamstown and Splott into her version of Tiger Bay) (2001). According to Reuschel and Hurni (2011:296), such authorial "transformations" fall short of locations being fully "invented", that is, those settings that have been constructed "within a familiar geographical reality".[1] It was clear, therefore, that Piatti and Hurni's sub-categories of fictional settings could be useful to impose *some* sort of control and organization on the complexity of a book's range of mappable spaces, but this control did appear to be somewhat arbitrary (where, for example, did the border between a "transformed" place and an "invented" place exist, and how do you know that you have crossed it?). It was nevertheless possible to identify, sift, and filter, and thus broadly (although not unequivocally) categorize different locations within this typology, with the purpose (for example) of analyzing how authors used different types of location to forward their story. For the *Literary Atlas* however, although an emphasis on the ways an author transformed a setting, or fully invented it, was relevant to our interests, it appeared to marginalize any focus on the reading experience and the creation of a "read-reality" that could emerge from a plotted approach. The approach taken by the *European Atlas* appeared too distant in its nature, with reading (especially by the "disembodied" modern academic) reduced to an initial scientific "splaying" (after Doel, 2018, see Chapter 2) and then subsequent analysis, rather than the identification of connections, and active synthesis.

It was clear, therefore, that there was a significant difference in approach and motivation between the *Literary* and *European* atlases. However, we remained aware that we were yet to fully engage with how we might identify our own extractions from the text. What was to be our "unit of analysis" (following Moretti, 2005, see Chapter 3)? How would we identify our key locations (however we were to term them), and how could we "cut up" our novels in order to punctuate (so to speak) our extractions? In order to help us come to a decision, we experimented with the opening pages of one of our shortlisted novels: *Aberystwyth Mon Amour* (Pryce, 2001). We constructed a worksheet which catalogued specific variables, namely: "Entry Number" (our first, second, third, etc. identified place within the text), the "Page Number of Entry" (from the specific edition of the novel we were working from), the location noted by the author, and the GPS co-ordinates of that location in order that we could map it (in the same way as outlined in Chapter 4). We then entered a short "co-text" from the novel, in other words, an extraction which would help to position and interpret how (and where) the author was using this location.

The first explicitly named location in *Aberystwyth Mon Amour* was "Canticle Street". This was an important location for the plot, and also for our plotted approach, as at first glance it was uncertain exactly where this location might be.[2] The *Literary Atlas* team was not the only readers to have encountered this mystery, as another, "Angela from Chepstow", remarked on a BBC online message board:

> I have just finished reading the third book by Malcolm Pryce. What brilliant reads. However one thing is bugging me. He makes several references to actual street names – Eastgate, Pier Street, Bridge Street etc – but he also mentions on several occasions a street called Canticle Street. Has there ever been or is there in Aberystwyth a street by this name?
>
> (in BBC, 2012, also cited in Anderson, 2014:200)

The author, Malcolm Pryce, responded to Angela in person, who then posted the following:

> [With reference to my question] I have just had a reply from Malcolm Pryce himself: "Hi Angela. Yeah, Canticle Street. Sorry you couldn't find it – there are a number of errors & inaccuracies in the real Aberystwyth. I've written to the mayor but it's no use. I'd stick to the books if I were you:)".
>
> (in BBC, 2012, also cited in Anderson, 2014:201)

As we can see from this exchange, although there is a "real" place called Aberystwyth, there isn't actually a Canticle Street in that "real" place. If we were to employ Reuschel and Hurni's typology of categorical relations (earlier), one could argue that Canticle Street is therefore an "invented" place (i.e. a setting "invented within a familiar geographical reality"), and at this stage, if we do not draw upon our tacit knowledge of Wales, we may even be tempted to describe Aberystwyth as an "invented" place too (due to it being composed up until this moment at least solely by this made-up place). Within one sentence, therefore, despite the apparent logic and order being imposed upon Pryce's literary geography by the *European Atlas*' typology of categorical relations, its limits are stretched and possibly exceeded.[3] Echoing the lessons learned through the cultural current across the social sciences (see Chapter 3), excising context from the co-text risks dis- or misplacing key information which could inform the nature of the place noted by the author. The imposition of categories may give the impression of omniscience in decision-making, and offer the guarantee of definitive outcome, but the lack of knowledge and the omnipresence of value judgement nevertheless continue to define the process. The process of categorization is not therefore a simple, scientific, or replicable process, but a thoroughly subjective and embodied one: it was clear that different "punctuations" in the co-text would lead to different categorizations.

Just as it was clear that how the user "punctuates" their reading of a (co-)text had important consequences for how a location was understood, so too did the type of reading a user brought to their extract. As has been discussed in Chapters 1 and 3, reading is not an essentialized activity, and with differing emphasis, new insights can be gained from the reading of the same co-text. This is exemplified with further reference to Canticle Street. If the co-text is limited to the direct sentence in which this location was first mentioned, the reader (as analyst) would know the following:

> I spent the years before the deluge operating out of an office on Canticle Street, above the Orthopaedic Boot shop.
>
> (Pryce, 2001:1)

As we have seen, from this co-text, the reader/analyst would not be certain what type of location Canticle Street was, or where it was (having no geospace to connect it to, or broader references from which to locate it). In short, the reader/analyst (or indeed A.I. algorithm) would not be able to categorize or locate this place.

However, if we moved beyond a modern analytical frame and towards a reading which was sensitive to how places are at once dramatic and geographic, it would be possible to gain much from this co-text. From this perspective, we could bring our own personal and professional "architecture of imagination" (after Larsen, 2018:168) to augment and interpret the co-text's c(l)ues (see Chapter 1). As readers, we would know that there was a "deluge" in this area (which may suggest that some sort of epic meteorological event had occurred, and given the "canticle" allusion, i.e. a reference to Christian liturgy, perhaps one that was even biblical in its scale or nature). We would also know that Canticle Street is a street where offices are situated, confirming to us that Aberystwyth, and this particular neighbourhood, is an urban place, but perhaps not an area that is ultra-modern or cutting edge in its modernity (again influenced in part by the reference to religion in its name, and a more general assumption (or even prejudice) that such sites may be generally located within more traditional, perhaps even old-fashioned areas of established Welsh towns). Such association is compounded by the office's location above a shop (specifically above an Orthopaedic Boot Shop). This suggests that the office is a small, perhaps down-at-heel operation (as it cannot afford to be located in a prime location), and to emphasize this point, is located in a street where other (very) niche outfitters sell their wares. Thus, if we adopt a plotted approach to this co-text, harnessing our tacit knowledge and "architecture of imagination" (Larsen, 2018:168), although this may not help the mapper locate Canticle Street specifically, it does help to conjure a rich "plotted" or "read-reality" for this location.

We can see therefore that the scoping decisions involved in punctuating co-text and choosing specific reading approaches can fundamentally influence the understanding, and locating, of the crafted words of an author. In order to help locate Canticle Street more specifically, the *Literary Atlas* team combined these two insights, expanding the extraction of co-text to its surrounding paragraph, and adopting a plotted reading of this extended section. This first sentence of this broader extraction read as follows:

> I spent the years before the deluge operating out of an office on Canticle Street above the Orthopaedic Boot Shop. And you know what that means: take two lefts outside the door and you were on the Old Prom.
>
> (2001:1)

When this "co-text" was extended to two sentences, more information was given to position Canticle Street. With this data, the reader could look to an existing cartography, in print or online, and see if there actually was (or is) a(n Old) Prom in Aberystwyth, and reverse the movements to see what streets might be suitable candidates for Canticle Street. Google Street View was used to see if the locale of each street broadly resonated with the description read on the page.[4] Through this investigation, one could suggest that Canticle Street is likely to be masquerading as Corporation Street or Portland Street. Later in the series, Louis Knight the detective (and narrator) decides to move offices, this time to a street that coincides with the cartographies of more accepted map-makers, as is described in *The Unbearable Lightness of Being in Aberystwyth*:

> "... Your office is in Canticle Street isn't it?"
> "Used to be, we've moved to Stryd-y-Popty."
> Today was the first official day of business in the new office and my feet hesitated for a second outside the old one but I forced them on up Canticle Street to take the unfamiliar left at the top into 221B Stryd-y-Popty. The street where the Poptys live, whatever they were. Judging by the windows, popty was Welsh for twitching net curtain. As with the old office, I entered through a door at the side and climbed up stairs coated in a carpet thinner than gossamer. There were two flats at the top – 1a and 1b. I hadn't met the occupier of 1a but judging by the wine bottles in the bin outside it was probably a student. Or the Mayor.
>
> (Pryce, 2005:22)

Complete with a tip of the fedora to Conan Doyle (presumably discreetly returned by a doff of the deer stalker), Pryce moves Louis' office to 221 Popty Street (Cymraeg for baker). This location can be found on conventional cartography, and if Louis takes the "unfamiliar left at the top" (see earlier) of Canticle Street to arrive there, this suggests the latter is actually Portland Street.

From this process, it was clear that the process of extracting and reading co-texts can directly influence not simply how places could be categorized (in the *European Atlas* approach) but also how they may be located for mapping more generally. As we saw in relation to the mapping candidate locations in Chapter 4, excising minimal co-texts risked displacing or misplacing locations. It was also clear that the widening of the extraction would help create a more vivid plotted or "read-reality".

Situating Sospan and the *Literary Atlas* approach

We experienced the benefits of choosing an extended extraction to inform a place's location when broadening a further example from Pryce's novel: Sospan's Ice Cream stall.

> Entry 4, Page 3, Place: "Sospan's Ice Cream Stall", Location: unknown.
> Co-sentence: "Normally when I made my stroll shortly before 9am I would see a few hanging around at Sospan's ice-cream stall."

When the reading of this co-text is attuned to the co-ingredience of drama and geography, little information can be gleaned to help detect the location of Sospan's Ice Cream stall. However, its crafted words do compel the reader to expand their understanding of the place by associating it with their own set of tacit knowledge and corresponding geographies – they create their own "read-reality" of this location. For example, we know there is an ice cream stall; this is not an established shop, but a stall, just for ice cream. We therefore know its scale, and since it is Sospan's (a unique name, and not Sospan's and Sons, or Sospan's Inc, etc.), it is perhaps not a large-scale franchise, but a low-key, one-person operation. Without knowing its location, but informed by the location of Louis' agency near the prom, we may anticipate it is located near the sea, and, given the vicissitudes of British weather, often exposed to inclement conditions. These crafted words also give further information about the nature of the place more generally; Louis Knight (the narrator) is strolling to Sospan's, he is not simply walking, or trudging, or marching, he is lazily "wayfaring" with a loose destination in mind (after Ingold, 2007). Normalized into his morning routine Louis is taking the air, passing the time, and without a strict deadline to meet. As this is happening at a time of day when many start work, this indicates to the reader that this is Aberystwyth's equivalence of the rush hour, which in turn suggests a local culture at odds with any metropolitan noise, stress, and associated hubbub.

If we extend the extraction to the surrounding paragraph, the co-text would reveal to us more about Sospan's stall.

> The thing I remember most about it was walking the entire length of the Prom that morning and not seeing a Druid. Normally when I made my stroll shortly before 9am I would see a few hanging around at Sospan's ice-cream stall, preening themselves in their sharp Swansea suits and teardrop aviator shades.
>
> (Pryce, 2001:3)

From this extraction, we know immediately that Sospan's is located on the prom (i.e. approximately co-ordinates of 52.421209 N, -4.084664 W), but we also know more about the prom itself. The day of Louis' stroll was unusual due to the absence of "Druids" on the prom – suggesting that normally there would be at least "a few". Given the novelty of such people existing in our (the *Literary Atlas* team's) everyday life, Louis' experience piques interest with respect to who these Druids might be, and what they may be doing in Aberystwyth. We do know, however, what they would normally be doing on the "prom": "hanging out at Sospan's". With this information, we now know that the ice cream stall functions akin to a diner, or drive through, a coffee shop, or a street corner: it is a place to "hang out", to see and be seen, it is a territory to be claimed, at least at this time of day, by the Druids. We now know, therefore, a bit more about Druids too – they seem like some sort of gang, who wish to be seen to be owning this place – perhaps suggesting a quasi-mafioso power over Aberystwyth. Such suggestions are emphasized with reference to their sartorial choices, suits of status, and assumed style, elevating Swansea in our "read-reality" to a place of rarified tailoring, and Druids to (wannabe?) Hollywood-style stars whose future is so bright they've got to wear (aviator) shades, even in the early morning. Through this detailing, druids at once emerge as Al Capone or M/maverick style hoodlums, and Louis as our underdog hero.

Beyond settings

Such generative engagements with short and more extended co-texts reinforced a number of insights about the approach of the *Literary Atlas* project. Our philosophical disposition, influenced by the relational sea change across the social sciences, and the currents of culture, space, cartography, and creativity that mobilize from it, produced an insistence that any decision to define locations as simply "settings" would be overly reductive. These engagements emphasized to us that all locations, at their point of being mentioned in the novel, have a time, have a culture, and have a geography, and when read this way, tell us something about not simply the place but also something about the drama. These locations are not therefore simply settings, but are better understood as points of the "plot" (in its broadest sense); in other words, they are "plot-points", intrinsically both dramatic and geographical.

Yet, what was also clear from these engagements was that the significance of any "plot-point" is not intrinsic to the text as the author crafted it. Rather it is interpreted by the reader, their broader knowledge, and the corresponding geographies they bring to inform it. The traction and significance of a plot-point therefore emerges through this process of creative definition. With this in mind, whether this plot-point is "invented" or "transformed" from an author's point of view, remains a fundamental part of the construction: it poses questions with regard to why the author chose this location, why did they keep it "real", or decide to reinvent it, in a particular way? It was vital that this part of the assemblage of literary geographies was included in our mapping process, yet what was also clear was this was only part of the assemblage. It was through the reading of these components, a particular reading that was sensitive to plot, that a fuller picture of this assemblage was gained. It was vital therefore that *Literary Atlas* should include all these components as part of its literary geographies.

So through a plotted reading of the novel we wanted to identify its "plot-points", contextualize these so we could map them as accurately as possible, and note appropriate extractions and subsequent interpretations that would create an iteration of this read-reality. To this end, the *Literary Atlas* team confirmed its decision not to simply data-mine in the distant mapping tradition, but to adopt a plotted reading for all the shortlisted novels. As we did so, we would note all and every external[5] location – or plot-point – we identified. In line with the working definition of literary geographies (outlined in Chapter 3), we were committed to remain critical and rigorous in this identification, but we were also informed by the current of creativity (see also Chapter 3): although our extractions could be predicted (identifying locations cited in the text), their interpretation and meaning could not necessarily be predictable or replicated, and our interpretations were not essential. Our reading and synthesis were therefore more directorial or curatorial in nature – we identified extractions, and varied our reading and pacing to pause, detail, interpret, and connect in line with how the crafted words caught and compelled us to do so. Within the Team, we didn't always agree, often argued, and our multiple insights and imaginative connections produced plural readings, and through so doing, our detailed units of analysis emerged.

Mapping plot-points

As we have seen in the opening of this chapter, in order to answer questions emergent from the spatial current, such as "where do stories *take place*" (see Anderson, 2021; Reuschel and Hurni, 2011:293), it is possible to convert one form of literary cartography (a book) into an abstracted bibliographic catalogue, and then translate this into a different visual form. For our identified plot-points, we decided to convert and visualize in a number of ways. To create a "baseline" map of our novels' plot-points, from which we could further augment and "deepen" our graphic representations (see Chapter 6), we set out to replicate the "distant" stratification we adopted in Chapter 4. With a catalogue of each novel's plot-points (with each location labelled, given co-ordinates, and an edited precis of its location and import), we presented these in aggregate for all our shortlisted novels. We visualized these on both a scrollable globe (www.literaryatlas.wales/en/maps/globe/), and (despite its acknowledged colonial undertones, the nevertheless familiar) Mercator projection (www.literaryatlas.wales/en/maps/locations/ see Figure 5.2).[6] Through these digital visualizations, users could scroll, identify, and zoom in scale to plot-points that interested them initially due to their cartographic location.

Such visual representation suggested that mapping offered the possibility of a new way of becoming engaged in the shortlisted novels, and perhaps even the wider canon of Welsh writing in English. Even if users knew little about this literature, they could nevertheless be stimulated to engage with our graphic catalogue through identification with existing places of interest (e.g. their home town, or even street, or familiar holiday locations) not simply in Wales, but across the globe. The précised interpretation of an associated plot-point may further stimulate interest in the plot, prompting engagement with the novel (or further maps of the *Literary Atlas*, see Chapters 6 and 7), or the place itself (see Chapter 11). The potential for mapping to generate broader cultural interest (in localities and their history, the

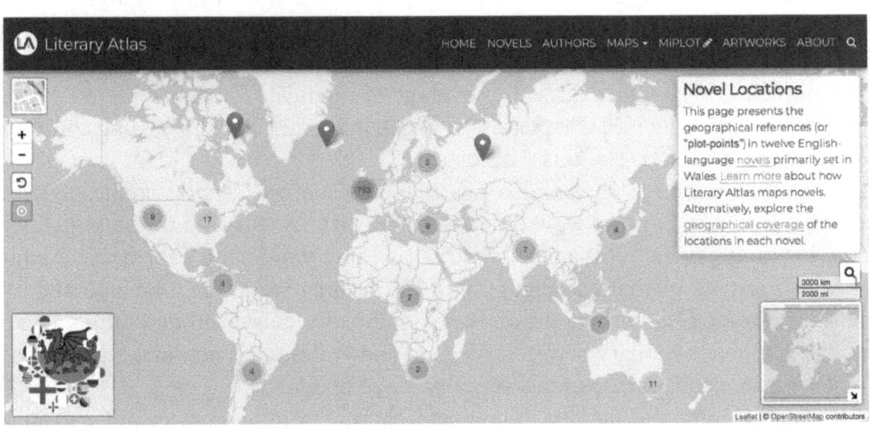

Figure 5.1 Mercator projection indicating locations of all plot-points from shortlisted novels for *Literary Atlas*.

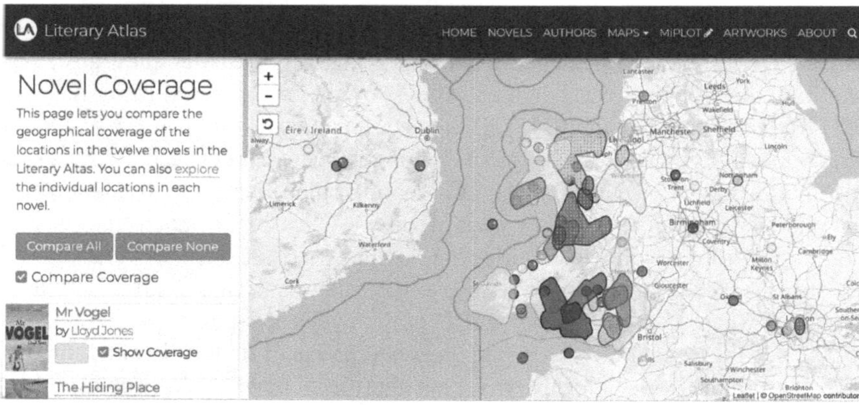

Figure 5.2 Map indicating regional distributions of plot-points for all shortlisted novels for *Literary Atlas*.

literature of Wales, and potentially even subsequent tourism to plot-points) was directly highlighted through the next step in graphic projection.

Co-ordinating cartographies

From this global and aggregated scale, we then visualized plot-points with respect to the novel they originated from. At this "scale", we experimented with a co-ordinate-based cartography. Akin to a choropleth map, these cartographies were thematic maps used to represent co-ordinates or other statistical data, and often used colours to indicate distributions and densities.[7] We used this type of map to indicate the key regions, predominantly in Wales, that each novel took place (see www.literaryatlas.wales/en/maps/coverage/ see Figure 5.3). By looking at all our novels together, we could get a sense of the total coverage of plot-points across Wales (and could therefore reflect on the appropriateness of our shortlisting decisions). The total coverage map mimicked to a broad degree the coverage of candidate locations offered by the catalogic cartography (see Chapter 4), so in general terms, our shortlisting could be seen to be representative of this database.

From this mapping, it was also possible to visualize the plot-points of each novel in particular. In this way, the user could identify what might be popularly seen, for example, as "William's Country" (in other words, the geographical coverage of *Border Country*, see Figure 5.4). Confirming what might be assumed by the title, *Border Country*'s plot-points predominantly cluster around Offa's Dyke and the southern boundary between England and Wales, with nominal mentions of Pembrokeshire and, perhaps most significantly, London.

We also created the facility to compare one novel to another in order to juxtapose how their plots may be shared (see below for geographically overlapping novels *Aberystwyth Mon Amour* (2001) and Niall Griffiths' *Sheepshagger* (2001)). This prompted possibilities of our read-realities becoming more complex, with plot-points from different novels overlapping in a user's "read-reality".

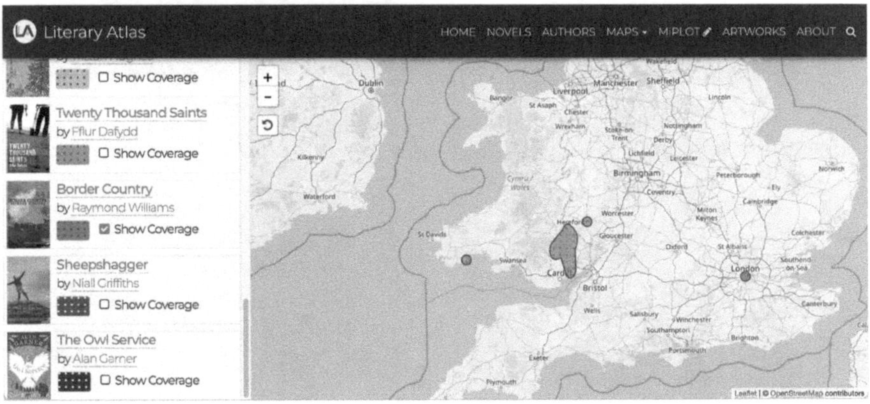

Figure 5.3 Map indicating regional distributions for plot-points for William's *Border Country*.

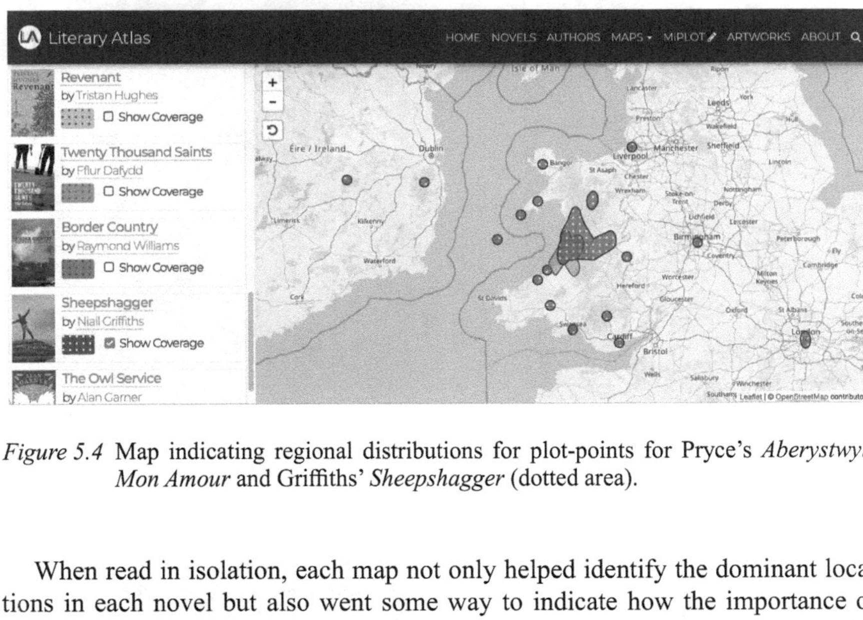

Figure 5.4 Map indicating regional distributions for plot-points for Pryce's *Aberystwyth Mon Amour* and Griffiths' *Sheepshagger* (dotted area).

When read in isolation, each map not only helped identify the dominant locations in each novel but also went some way to indicate how the importance of relative distance between key plot-point groupings, thereby suggesting further significance to their respective plot. For example, the representation of plot-points for Fflur Dafydd's *Twenty Thousand Saints* demonstrates not simply the key locations of Ynys Enlli/Bardsey Island in the north-west coast of Wales, but also its peripherality to the main "body" of Wales, to key population centres, and indeed to the capital Cardiff. This marginality and relative distance becomes a material and metaphorical theme in the novel, underpinning Delyth and Viv's identification of Ynys Enlli as the ideal location for their initial escape from the mainland's political climate, and their dislocation from the subsequent changes that were set in motion.

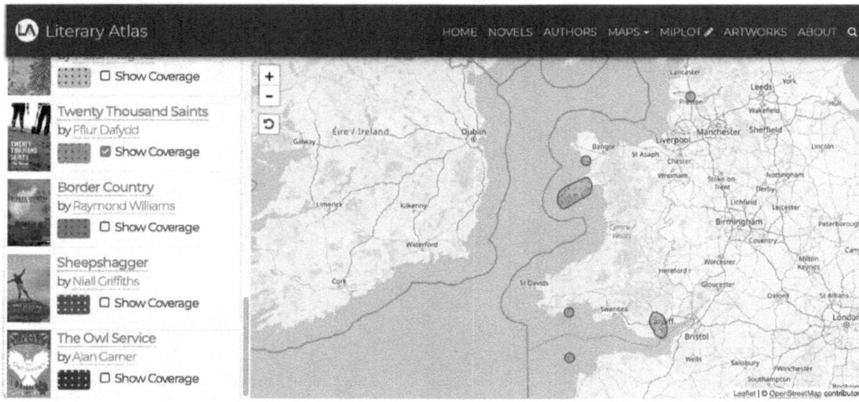

Figure 5.5 Map indicating regional distributions for plot-points for Dafydd's *Twenty Thousand Saints*.

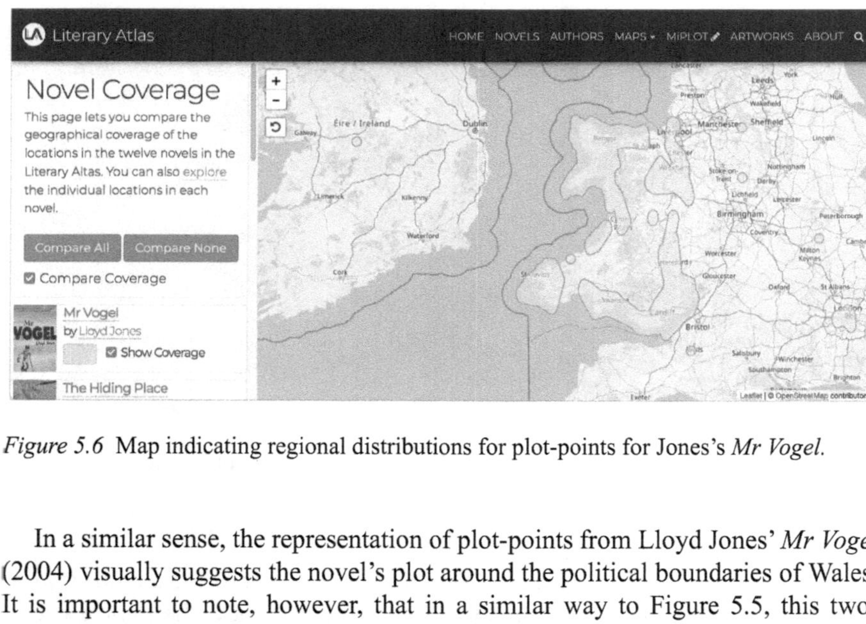

Figure 5.6 Map indicating regional distributions for plot-points for Jones's *Mr Vogel*.

In a similar sense, the representation of plot-points from Lloyd Jones' *Mr Vogel* (2004) visually suggests the novel's plot around the political boundaries of Wales. It is important to note, however, that in a similar way to Figure 5.5, this two-dimensional visual representation does not give the reader any sense of the timeline or order of the novel's movement, and the significance this may have (or not) to the plot. In this sense, the potential remains to add new dimensions to these maps to further contextualize and represent the plotted reading approach.

Conclusions: centring on a plotted approach

As we have seen in this chapter, when beginning to position the approach of the *Literary Atlas of Wales*, its emergence under the influence of the relational sea change in the social sciences, arts, and humanities meant that it was fundamentally different in philosophy and approach to the *Literary Atlas of Europe*. Although it produced distant maps (see Chapter 4), the data that produced these maps, and the ways in which these maps were used, were fundamentally rooted in a plotted approach. In this approach, the academic is not scripted in the role of scientific analyst, but is rather framed as a critical and rigorous reader, who makes a series of active decisions on which locations may be identified (for example, as external or internal geographies), the size of co-text used to inform these locations (including how these co-texts may be punctuated or otherwise interrupted to emphasize particular meanings), and how these may be interpreted and connected to a broader awareness of these literary geographies. Inevitably, therefore, a consequence of a plotted reading is that it reflects in part what an individual (or team of) reader(s) brings to their engagement – the read-reality produced is a relational process, emergent through this practice. Through their series of aggregated decisions, the role of the academic has moved away from being simply an analyst, and towards the position of being an auteur, with their own backgrounds, insights, and objectives directly influencing the direction, content, and framing of the maps produced.

Following the consolidation of these insights, the *Literary Atlas* team progressed to identify their own version of key locations from our selected novels. From a plotted reading of all novels, we identified key "plot-points" or co-ingredient sites of drama and geography, informed by the broader novel, as well as the reader's wider literacy of both. Once visualized on global projections, these "plot-points" became the basis for distribution to outline "regions" of literary geography (similar to what Chapter 11 will discuss in relation to conventional literary tourism activity). These regional maps served to produce new insights about the geographic distribution of the novel and the consequence of this for its drama. In so doing, these maps raised the importance of a temporal dimension, about how one plot-point may lead to another, and the concomitant opportunity (and perhaps even necessity) to further contextualize and map any series of plot-points in further innovative ways. Could different visualizations further "deepen" the potential promised by the plotted approach? This question and its answers are engaged with in detail in Chapter 6.

Notes

1 For the *Literary Atlas*, it was possible to identify "invented" locations with reference, for example, to Sospan's Ice Cream stall on the "geospace" of the prom, in Malcolm Pryce's *Aberystwyth Mon Amour* (2001).

2 Furthermore, it was also unclear what type of "setting" this may be in relation to Reuschel and Hurni's categorical relations.

3 Indeed, in the co-text extract in which Canticle Street is referred to ("I spent the years before the deluge operating out of an office on Canticle Street, above the Orthopaedic Boot shop" (2001:1)), Canticle Street is not a location that the characters are "in" at this moment of the story, the (as yet unknown) narrator is informing the reader where they were at some point in the past – Canticle Street was their "setting" at that time, and the reader does not yet know whether they are still located there, or indeed whether the "setting" still exists at all. The possibility is raised therefore of categorical relations between literature and place changing, depending on what co-text is chosen.

For those readers who have completed the novel, or engaged with others in the series, they will know more about the location and importance of Canticle Street to the story. They will know it still exists for the protagonists, and it becomes a key location. They will also know (as the exchange between "Angela from Chepstow" and Pryce implies) that the Aberystwyth of the novel has few "invented" places within it. But from within the limits of co-text extract, no definitive insight for categorization can yet be gained. Decisions in respect to context therefore remain key to informed categorization within this typology.

4 And why not give it a try: www.literaryatlas.wales/en/novels/aberystwyth-mon-amour/explore/#section=canticle-capers&plotline=real-aberystwyth&figure=*).#*

5 If, for example, a particular "living room" was noted, we would create a plot-point for the house, but not the domestic specifics. The mapping of domestic geographies would be possible, and offer many interesting insights for particular novels, but it was not the objective of the *Literary Atlas*.

6 Alongside these projections, we offered a cluster of flag representations, with respective flags proportionate to number of plot-points located within that nation.

7 At this stage, we chose not to map the frequency of place "hits" – we worked on the assumption that significance is not always indicated by frequency, but by the context in which a location is used as a plot-point. This insight we returned to in Chapter 6 when constructing plotlines.

6 Locating the *Literary Atlas*

Deepening plot-points

Influenced by the cartographic and creative currents (see Chapter 3) and informed by the project's functional interdisciplinarity (see Chapter 2), *Literary Atlas* was keen to foster the potential promised by bringing together different knowledge "fragments" to create innovative mapping representations (see Chapters 4 and 5). Maps were now no longer considered to be simply knowledge banks, but rather resources to stimulate curiosity, creativity, and cultural engagement. As Chapter 4 has discussed, the *Literary Atlas* team had undertaken a plotted reading of all the shortlisted novels, and created a log of every existing plot-point. These had been connected to a set of geographical co-ordinates, and then, drawing on our literacy and knowledge, we had attributed to them our own interpretation of their status and significance.

From these initial developments, it was clear that the plotted approach to literary geographies offered a different philosophical and practical take when compared to already-established atlas projects, as well as suggesting further possibilities for innovation. Through making explicit the practice of plotted reading, it was clear how the crafted words of the author could combine with a reader's (in this case the *Literary Atlas* team's) own corresponding geographies[1] and create a stimulating and centripetal "read-reality" (or borderland, after Pullman, see Chapter 1). Informed by this approach, we were interested in how the potential of this process could be harnessed further. This chapter charts this process of investigation, and begins by explicitly acknowledging how "deep mapping" (see also Chapter 3) could realize the aims of the project in key ways. As we will see in this chapter, deep mapping is a practice that can help to assemble new perspectives, connections, and dimensions to not only the plot-points of our authors but also the corresponding geographers of our reading. Deep maps can help produce what Harris (2015) terms new "spatial narratives", or more specifically, new plotlines through the borderlands of our selected novels.

Diving into a cartography of depth

As we have seen in Chapter 3, deep mapping has been a response to a cartographic current which has challenged the very idea of what a map is or could be. Deep mapping, as its name suggests, does not seek to reproduce an "artificial" (see Moretti,

DOI: 10.4324/9781003321248-6

2005) or "thin" description of a place (Harris, 2015:30), but echoing anthropological impulses for "thick descriptions" (after Broyard, in Least Heat-Moon, 1999:8) seeks to bring together a rich array of "details of life until they begin to coagulate . . . in an interpretation" (Least Heat-Moon, 1999). Deep mapping therefore seeks to produce a multi-layered mix of details about a location, celebrating perspectives from often contradictory cultural standpoints and different historical time frames. Perhaps unsurprisingly, given their shared emergence through the cartographic current (see Chapter 3), deep mapping and the plotted approach are broadly resonant in their nature. Both begin with an acknowledgement of and commitment to a relational constitution of the world (see Chapter 3, but see also Biggs (in Hawkins, 2021:42) with particular reference to deep mapping), and an amodern celebration of difference and multiplicity.

The plotted reading approach adopted by the *Literary Atlas* undertakes a practice which can be seen as broadly similar in process and intent to the act of deep mapping. In order to create a plotted reading, it is necessary to position oneself as an active interpreter of the crafted words of an author, bringing to their plot a range of often subjective and possibly contradictory associations and correspondences. In order to deep map, as Naramore Maher tells us, it is necessary to: "limn a matrix of communal and personal memory, competing histories, and polyvocal narrative" (2014:XIV). To deep map, the cartographer must bring together "a multiplicity of sources [,] perspectives [, and] experiences" of a place (Naramore Maher, 2014:132). As Harris suggests, these sources, perspectives, and experiences could involve a range of media, including, for example, "moving images, virtual creations, aerial imagery, Google Streetscene, text, sound, and personal messaging" (after Harris, 2015:45, but see also McLucas (2014), and Pearson and Shank (2001)). Aligned with the spatial current (see Chapter 3), this "cartography of depth" (Roberts, 2016:6) does not seek to produce a definitive or "objective" framing of a place, but rather contributes to an inherently unstable "negotiated conversation" about it (after Bodenhamer et al., 2015:21). Such mapping perpetually re-(and de-)composes, actively responding to "new data, new perspectives, and new insights" (Bodenhamer et al., 2015:4).

In this way, the plotted reading approach could both benefit from and add to the practice of deep mapping. It can be seen, firstly, that the plotted approach and deep mapping both appear to be part of the wider project which seeks to "put the context back" to any "object" of study (see Chapter 3). In contrast to the construction of distant maps which, as Chapter 3 explains, could be summarized by a process which "abstracts, reduces, and extracts", deep mapping and plotted reading could be summarized as processes which "identify, relate, and expand". This difference in process directly relates to differences in intent; where distant maps seek to construct simplified "data spaces" (Benedikt, 1991:135) which deny or reduce the "uncertainty" and ambiguity of the world (after Reuschel et al., 2013), deep mapping and plotted reading seek to enable "data spaces" to emerge which acknowledge complexity and ambiguity. Deep mapping can therefore be seen as a practice that can help to further "put the context back" to *Literary Atlas*'s object of study, which in turn could expand our knowledge of any location; *Literary Atlas* however, would not simply be producing a deep cartography of place in the deep mapping tradition, but supplementing this practice with a deep cartography of plot.

Deep plots

In order to deepen the plotted readings of our selected novels, the *Literary Atlas* team began to accumulate a diverse set of resources which could help produce new contexts for and perspectives of our plot-points. We began with secondary data, and the identification of digital maps representing physical and political boundaries of Wales. From these bases, it was possible to impose overlays of different historical and social data sets (for example, from the time period each novel was set, but also when it was written and published, as well as for key moments when its reception may have changed over time). We then accessed academic, cultural, or authorial commentaries occurring in these time periods, as well as compiling "deep" localities data, including Census, Labour Force Surveys, Welsh Health Surveys, and bespoke surveys undertaken by local authorities which produced, for example, deprivation statistics. We also considered a range of other visual representations of our plot-points, including still and moving images of locations, in a variety of genres (e.g. films or documentaries), and from a variety of media sources (e.g. mainstream, alternative, or social).

Allied to these secondary sources, it was also possible for the *Literary Atlas* team to add meaning and interpretation of our plot-points through engaging in primary data collection, principally through conducting author and expert interviews. Although we were aware that some scholars working in the arts and humanities may not consider the insight of an author relevant to a discussion of a novel,[2] influenced by the cultural current (see Chapter 3) and specifically the impulse to consider the intentions of an actor, their perceived success, and their awareness of the context in which they were writing, we found the possibility of engaging with authors to be both desirable and exciting. In order to operationalize this possibility, we chose from a suite of qualitative methods,[3] more particularly, interviews. As Ackroyd and Hughes put it, interviews can be understood as "encounters between a researcher and a respondent in which the latter is asked a series of questions relevant to the subject of research" (1981:66).[4] The interviews we conducted were "semi-structured" in nature, based initially around key questions generated from our plotted readings (see Chapter 8), but also open to new avenues prompted by our interviewee's responses. This capacity for flexibility was especially important given our acknowledgement of the role of "plot" to our methodology.

As noted in Chapter 3, the spatial current has had effects on the conceptual and methodological orientation of many social science, art, and humanities disciplines. As such, scholars have acknowledged that the geographical place of method is just as important to knowledge-generation practices as the "social" dimension (with the latter, for example, influencing the need for reflexivity with regard to the positionality of researcher and respondent, see, for example, Arskey and Knight, 1999; Sin, 2003). Following the spatial current, research method is configured less of a "dialogue" between the researcher and researched, and more of a "polylogue" (after Anderson et al., 2010) between researcher, researched, *and* the place of method (see also Anderson and Jones, 2009). With this in mind, the *Literary Atlas* team actively considered how the plot-points already identified could be integrated into our methodological engagement with authors or experts. Wherever possible, we attempted to interview individuals in a location that corresponded to the plots under discussion, so their dramatic and geographical significance could act as cues for directing conversation and prompting associated knowledge.

Influenced by the spatial current, we were also aware that the *mobility* of method offered the potential to access insights into storytelling and spatial narrative decisions. In recent decades, many social scientists have employed interviews "on the move", using particular practices such as "talking whilst walking" to enhance the interactions between respondents, interviewers, and places they move through (see, for example, Anderson, 2004). Notwithstanding the importance of mobility issues, it was noted that harnessing the peripatetic affordances of the "conditions, qualities and rhythms of the body in motion" (Jarvis, 1997:IX) could help excavate associations created by individuals to our plot-points. We therefore employed mobile methods to this end, walking, for example, with Fflur Dafydd on the Llyn Peninsula (viewing Ynys Enlli/Bardsey Island off its coast), getting lost in the hinterland hills of Ceredigion with Niall Griffiths, and perambulating North Wales with Lloyd Jones. We recorded these encounters between author and place using audio, photo, and film technologies, endeavouring to capture atmospheres, emotions, and reflections as well as the ideas motivating their spatial storytelling.

Thus, the components we identified to contribute to our deep mapping process began with plot-points from the novel in question, but combined with corresponding others from the *Literary Atlas'* wider catalogue. From this basis, the deep maps were layered with historical and contemporary resources in the following areas: (digital) maps (which outlined physical and political borders along with a variety of social data sets); popular media resources (including still and moving images, freely available documentaries or films of the plot's locations and corresponding geographies, along with contemporary street view recordings); audio and visual material from mobile interviews with authors or experts (if the authors were deceased); and literary commentaries from academia and popular media. With these data sets compiled, the key issue was now how to appropriately present these to an audience.

Curating deep maps

Following the collection of the range of deep mapping materials noted earlier, it would have been possible for the *Literary Atlas* team to assemble these resources in an open digital archive and allow users the freedom to interrogate them as they wish. Each plot-point and corresponding component could "simply" be presented on an interactive map, and map-readers invited to follow their curiosity through this assembly, heading out on a metaphorical "way fare" (after Ingold, 2007) through locations, literatures, images, and sound recordings (etc.). Such a process would resonate with, and to some extent mimic the *Literary Atlas* team's own process of exploration, allowing users the possibility of "enchantment" by reacting to their own small discoveries and making their own connections; in the words of Woodyer and Geoghegan (2012), it would allow users to not be encompassed by an overriding analytical narrative, but rather be self-directed:

> tak[ing their] time to appreciate the small things and allow[ing . . . them] selves to be nourished by the energy their surprise affords.
>
> (Woodyer and Geoghegan, 2012:195)

Although the promises offered by open presentation were persuasive, they also risked confusion. Replicating the limitations of a "macroscopic" approach (i.e. the study of large or complex entities/processes), the imbroglio of data sources may make it difficult for the user to connect with them in any meaningful way. This is a problem that Eanes et al. (2020:8–9) have identified in relation to presenting deep maps in particular; they note, for example, that

> by themselves, . . . interactive deep maps [can] simply be chaotic, if intriguing, collections of multimedia observations, rendered over a mashup of [cartographic] data layers . . . In sufficient amounts, such unorganized spatial information, whether qualitative or quantitative, can become cognitively overwhelming.

In order to counter this risk, it was deemed important to not simply collect resources for deep geographies but to actively curate them. To the *Literary Atlas* team, it was clear that for some plot-points a particular assembly of deep mapping resources were more relevant and stimulating when compared to others, so we drew on our relative expertise to juxtapose these fragments in the hope that they would maximize the possibility of the user experiencing their own "moments of illumination" (echoing the process of Benjamin in Paris arcades, discussed in Cresswell, 2019, see also Chapter 4). In this way, our curation became oriented around compiling "an innovative, culturally nuanced and critically 'responsible' heuristic through which to encounter" our deep mapping materials (after Walford Davies, 2012:207).

Operating on the cusp of technological advances in augmented and virtual realities, and with the widespread public accessibility of an array of cartographic software, we remained sensitive to how the form and interactivity of visual media could be used to enhance the user experience of our deep mapping materials. Following both the visual ethnographic (Scarles, 2010) and poetic traditions (Smith, 2024), we also hoped that our use of media could add meaning to user engagement with our cartographic plots. We considered how formatting conventions of different media helped users to "make the transition from one scene or one idea to the next" (Hintz, 2020), and with this in mind, we similarly reflected on how both split screen formatting (used, for example, in sports reporting and news interviews) and digital map layering (employed in car and phone navigation systems) had become increasingly familiar in popular culture. It was considered possible, therefore, to present and juxtapose our data components on digital split screens, oriented around a combination of map layers, whilst also reducing any significant skill or literacy barriers to engagement. Within these broad parameters, we drew together basic insights about user engagement with websites,[5] with broader work in deep mapping more generally (Eanes et al., 2020; Wang and Ye, 2020) to reflect on how to bring together our range of sources and "best utilize multiple semiotic modes, such as the aural, visual, gestural, linguistic, and technological, to fulfil [our] objectives" (after DePalma and Poe Alexander, 2015:188). In short, we aimed to create an online resource with a clear and stimulating cartographic base, including temporal dimensions and public interactivity wherever possible, to maximize usability and minimize the risk of macroscopic "chaos" (see earlier). To realize this, we composed our deep maps in two key ways: through "synopsis" and through "plotline".

Synopsis and plotline

As its name suggests, the "synopsis" of each selected novel offered an overview of its plot to the *Literary Atlas* user. Each "synopsis" introduced the novel and was presented with a distribution map of the novel's plot-point coverage (see, for example, the synopsis for Trezza Azzopardi's *The Hiding Place* (see www.literaryatlas.wales/en/novels/the-hiding-place/ and below) as well as the synopsis for Alys Conran's *Pigeon* (www.literaryatlas.wales/en/novels/pigeon/ see below)). These synopses were combined with audio extracts (which can also be found on the website links listed) which also helped to introduce the novels (read by their author and/or the *Literary Atlas* team), along with historical and contemporary maps of the novel's plot, street view images, or other photos and films connected to key sites.

Synopsis 1 (text version): The Hiding Place

> *The Hiding Place* (2000) was Trezza Azzopardi's first published novel. It tells the story of the Gaucis, a half-Maltese, half-Welsh family living in Butetown – "Tiger Bay" – in Cardiff. The story is told through the first-person narrative voice of Dolores, the youngest of 6 daughters of Frankie Gauci, a Maltese immigrant, and his wife Mary. It is the late 1990s and Dolores has returned to Cardiff to attend the funeral of her mother and reconnect with her sisters. Her return to the locations of her childhood evokes vivid, traumatic memories and perspectives on her upbringing in the 1960s.

Main characters

The Hiding Place is narrated from the first-person perspective of Dolores Gauci, the youngest of six sisters. Dol, Marina, Celesta, Fran, Rose, and Luca are the daughters of Mary, who has moved to Cardiff from a small village in the Cynon Valley, and Frankie, a Maltese immigrant who has arrived to build a life for himself in Cardiff. Upon his arrival in Tiger Bay, Frankie meets Joe Medora, a small-time criminal working his way into a position of power within the world of organized crime. Frankie also befriends Salvatore Capanone, with whom he opens a café on Bute Street.

Location

The novel is set in Butetown – an area near the docklands area of Cardiff known colloquially as "Tiger Bay". Tiger Bay famously supported a rare kind of multicultural community due to its close proximity to Cardiff Docks, a busy port during the height of the Welsh coal industry in the late nineteenth and early twentieth centuries. The shipping industry brought in migrants from around the world to work and live in Cardiff, and the result was a richly multicultural community populated by people of many ethnicities, languages, and religions.

However, there were changes afoot in the 1960s, when many of the areas surrounding the Cardiff Docks were demolished to make way for new urban developments. This substantially changed the community. In an interview with Literary Atlas, Trezza Azzopardi recalls growing up at the time of this massive urban change:

> "So as I was living there, they were knocking down the streets behind our terrace and it was one of those odd early politics moments where you understand that things are happening outside of the world that you can't control. One of them was that there were loads of women saying, well we're going to go and lie down in front of the bulldozers, because they didn't want to be moved to a place that they didn't recognise, which was literally in fact only down the road. But all the same, you know, their houses were being knocked down."

Much of the novel takes place against the backdrop of these changes in the urban geography of Cardiff during the 1960s, and it captures the darker aspects of life in Tiger Bay at that time. When Delores returns to the area as an adult in the late 1990s, she finds Cardiff in the midst of a new phase of urban development. The novel explores the ways in which places, despite cosmetic regeneration and change, can still be charged with troubling memories.

Though *The Hiding Place* is clearly set in Cardiff, the novel occasionally distorts the real geography of the area. While some landmarks and road names are accurately described, others are moved and renamed for the purposes of the narrative. Azzopardi remarked on this during her interview with Literary Atlas:

> in terms of things like . . . one of the characters goes searching through the streets for another character, and those streets are actually in Adamsdown. Parts of them are in Splott, and one commentator has said, whoever thinks this book is set in the Docks doesn't know Cardiff 'cause actually it's set in Splott. So for me, it's as I remember it but for people who maybe live there now, who go and forensically look at it, there might be a slight distortion of the geography.

Synopsis 2 (text version): Pigeon

> *At the top of the hill they looked out at it. Their town. Its slow streets. Its shifting houses. The fidgeting gardens where cats trailed across fences, catching their small springtime prey, the poking chimneys, a couple of them smoking, even on a warm day. The town mumbled to itself beneath the deep clouds.*
>
> *(Conran, 2016:243)*

Pigeon is a story about the role of our everyday geographies – our home, our street, our town and its people – in forming who we are. It asks how we story ourselves in relation to the inherited cultures and histories of a place (or the combined "mumblings" as Conran refers to them), and how these may often be comforting but sometimes intensely claustrophobic.

Pigeon is a novel about growing up and learning to find a place for ourselves amongst these familiar cultures and geographies, and how it is often only when we leave, and then return, that we can do so on our own terms. In sum, the novel asks how do we build a place for ourselves in the world, on our own terms, in our own words.

Pigeon is set in an unnamed mining town in North Wales. The town lies on the periphery of the *Snowdonia National Park*, and its cultural landscape is dominated by the juxtaposition of the rugged beauty of this topographical surface and the quarries that mine the *geological riches of the slate* beneath.

The author Alys Conran grew up in this region, and has returned to it following a period abroad. This story of growing up, exile, and return finds a parallel in the themes dominating the novel, and in particular, the title character of Pigeon.

Main characters

The novel focuses primarily on two characters, Pigeon and Iola. These are two young people who are marginalized and dispossessed from the constraining familial and cultural structures of their town. By giving voice to these characters from the margins, Conran echoes other novels in the *Literary Atlas* (see *Amy Dillwyn* and *Niall Griffiths*), introducing readers to the perspectives of individuals who may be easily ignored, overlooked, or silenced by dominant sectors of society. The novel charts Pigeon's journey in finding his own voice and place.

"I was aware that I felt that the stories of young people who struggle in places like this were not maybe finding their way in North Wales, were not finding their way into books and I wanted to write about a boy like Pigeon, a boy from Pigeon's background and a boy with Pigeon's struggles. I think I also wanted to write against this, this stereotype that I'd found restrictive, of Wales where those kind of characters perhaps had no creativity or agency or imagination within their situation. So Pigeon does find himself at several points unable to change his situation but he's always a character that is absolutely full of ideas and really able to imagine his way out. I definitely wanted that" (Conran, in an interview with *Literary Atlas*).

Conran describes Pigeon and his apparent alienation from the town in which he lives, as follows:

> *So here, here's Pigeon again. Here, grey. Just a sketch the boy. His face is sallow. There's a snarl at his lips, and his shoulders are delicate as eggshells. Pigeon, here on the hill, wanders the pebble-dash, pebbled ash, scuffing his feet up the hill, and then up between the houses.*
>
> *(Conran, 2016:14)*

Iola is introduced to the reader in the following way:

> *She appeared. It was lola, come out of the wood like a genie, small with a pot belly over her skirt, knees as always covered in bruises and grazes, shoes undone, hair so light it's almost white.*
>
> *(Conran, 2016:6)*

Both these characters are defined to some extent by their relation to the world around them. They are both vulnerable and fragile, as if just emerged from a world of mythical nature into a hostile environment that will graze and bruise them into submission, if they don't get their scuffing blows in first. As we will see, both characters appear out of place in the urban environment of the fictionalized Bethesda, and more at home in the woods and hills surrounding the town. They find escape both in these areas and in their relationship. Both these sets of connections help form a sense of who they are, and what they may become.

He [Pigeon] needed her. Without lola, all his thoughts just made a black inky mess.

(Conran, 2016:34)

Each synopsis thus served as an introduction to the *Literary Atlas* team's plotted reading of the novel, yet the plot-points, their corresponding geographies, and their associated deep mapping materials remained potentially macroscopic and complex. As they stood, there was no mechanism, or "path", which could offer the user "not only a means of traversing" the multiplicity of deep mapping materials, but also ways of "feeling, being and knowing" them (after MacFarlane, 2013:24). The *Literary Atlas* team sought to provide this means of orientation through the curation of plotlines (see Anderson, 2014).

According to Anderson, plotlines are routes through the coming together of geographical locations, fictional narratives, author's viewpoints, and readers' experiences.[6] At their basic level, plotlines offer a new path, or line, through a series of plot-points in a novel. This line rarely follows the narrative chosen by the novel's author, but rather offers a new, curated journey prompted by the plotted reading process and following a thematic or otherwise salient new route. For the *Literary*

Atlas project, these new routes harnessed not simply the crafted words of the author and the plotted reading of its curators, but also the deep mapping materials compiled by the *Literary Atlas* team. Cast in this way, plotlines offered the potential to be a series of threads that collectively link the data provided by deep mapping in ways that ultimately produce meaning. In other words, it was hoped that plotlines could offer:

> a means for structuring the content of interactive deep maps in ways that are cognitively appealing to both mapmakers and map users . . . guid[ing] the user through the multiple perspectives and artifacts contained in the map, [and] allowing for coherent spatial experiences and arguments to emerge.
>
> (Eanes et al., 2020:9)

An example (text version) plotline, in this case the first of four charted for Azzopardi's *The Hiding Place* follows beneath (it can be found in its multimedia version here: www.literaryatlas.wales/en/novels/the-hiding-place/).

Plotline 1 (text version): Frankie's journey (from Azzopardi's *The Hiding Place*)

Cardiff Docks

Frankie Gauci's journey is similar to the many thousands of dockworkers and sailors who migrated to Tiger Bay during the late nineteenth and twentieth centuries: "seamen from the Tramp Trade who came to rest and stopped for good", as Dol describes them *(p. 24)*. He arrives as a sailor in search of further work: "Frankie knew what to do when he came into port: register, find a place to stay, then cut another passage on the sea" *(p. 40)*. Yet, like many others before him, he soon sets up a life in this multicultural district and becomes a resident. In fact, he has already heard of Tiger Bay as a place where many Maltese have arrived before him:

> *He hasn't chosen Cardiff by accident: he's heard there's a great clan of Maltese here, with more arriving every day. There is money to be made at sea, he's been told, and this is the place to spend it. Tiger Bay – the Valletta of Britain!*
>
> *(p. 42)*

Sliema, Valetta

Upon his arrival in Cardiff, one of the first things Frankie notices is the cold weather. This physical sensation reminds him of the warmth of his hometown of Sliema in Malta.

Frankie has never been so cold in his life. It's not the feeling you get on board ship, when the squall punches your face, stabs at your teeth, when your whole head is a sharp pain. . . . Nor is what he feels anything like home, where winters are short and February not so cruel. Frankie thinks of Sliema, of the sandy lane winding up to his village, with the sky a soft grey and the rain so fine it's hardly felt at all.

This cold is a slow ache; it makes your skin sore, it makes you want to crouch double.

(p. 39)

Seamen's Mission

Frankie's first port of call in Cardiff is the Seamen's Mission to register as a sailor. This is his first contact with Welsh people – though Frankie doesn't seem to understand that he is in fact in Wales, not England:

> *"Your first visit to Wales, is it?"*
>
> *and Frankie, who knew enough of what was being asked, nodded again and said,*
>
> *"Yes, first time in England".*
>
> *At last the man smiles at him.*
>
> *(p. 41)*

Frankie and Joe's room

Carlo Cross offers Frankie a cramped room in a basement flat in Butetown. The room is an oppressive space, which appears to match Frankie's sense of physical and emotional discomfort in his latest home:

> *Frankie sat beneath the window, dabbed at his nose with his handkerchief, studied his new home. The ceiling sloped down steeply in one corner, with a bed wedged in the gap below. The first night, he couldn't sleep for the traipse of feet on the stairs above his head, so close to his ear they seemed to be stumbling over his face.*
>
> *(p. 42)*

Frankie assumes that, given the smallness of the room, he is the only tenant. Yet he soon discovers that he is sharing the space with Joe Medora, a fellow Maltese migrant. It is this relationship that comes to define Frankie's life in Cardiff.

The Hayes and the market

Frankie and Joe soon strike up a friendship. Two young bachelors in a new city stroll around the place, taking in the sights and enjoying their freedom.

Their lives weave quietly together. They have a pattern: a late breakfast at The Hayes, followed by a stroll to the market where Frankie buys food. He handles and sniffs and scowls at the produce until the stall-owner loses patience, or Frankie gives up in disgust. (p. 46)

Frankie's interaction with these city spaces signals his growing confidence in his new home; no longer oppressed by the slanted ceiling of his room, Frankie now has the freedom of the city, even bartering with local stallholders at the market. Meanwhile, Joe is quietly establishing himself as a petty criminal, working small-time jobs:

> *While Frankie [spends time in the market], Joe goes for a talk with his "Boss" in the pub across the road. They often have an errand after this meeting – to collect a debt, drive a girl to a hotel, deliver a parcel: sometimes, all they have to do is sit on the street and watch. They are paid handsomely for what seems like nothing at all.*
>
> *(p. 46)*

Port of Call Café/Moonlight Club

Having settled in Tiger Bay, Frankie goes into business with his friend Salvatore. They open an establishment on Bute Street and share ownership. This begins life as a café, but later becomes a seedy, but rather more lucrative, bar. Frankie's connection with this space as the years go by symbolizes his wider mistakes and misfortunes in life. Initially, having built a life alongside his friend Joe, he inhabits Bute Street with a sense of swaggering and proprietorial confidence:

> *Frankie strolls past the Bute Street cafes, nodding now and again at a familiar face, or raising his hatted hand in a greeting. This is Frankie's patch. Most of the restaurants and cafes are owned and run by his friends: seamen from the Tramp Trade who came to rest and stopped for good.*
>
> *(p. 24)*

However, Frankie's confidence proves to be built on shaky ground. He isn't interested in running his café; too busy cultivating his gambling addiction, he lets Salvatore do the real work. This proves to be his downfall. Racked with rising gambling debts, Frankie loses his half-share in the business to Joe. This changes the power dynamic in Frankie's relationship with his old friend: now the relationship has become one between debtor and creditor. The office that Frankie once occupied becomes Joe's. It is now a space to which Frankie is summoned when Joe wishes to make him an offer he cannot refuse. Frankie's powerlessness in this space is performed through his uneasy body language:

> *Frankie eases his hold on the sweating doorknob, cutting the silence with a shriek of spring. . . . Frankie is all eyes, taking in his old den, now transformed into Medora Territory. . . . He shifts his weight from one foot to the other. Ilya grins suddenly, amused by the sight of my father's discomfort. They both know that Joe Medora could make him stand there all day. It is a battle which Frankie can't win; he must stay calm.*
>
> *(p. 36–42)*

Eventually, Frankie's misfortunes lead him to further criminal acts. Desperate to leave Tiger Bay, he raids the safe at business he once owned to fund his escape.

On the Athene

The last place we see Frankie is on the Athene, the ship on which he makes his escape. He lies back in his bunk and smokes, seemingly unrepentant.

My father knew what he had done. He sailed out at dawn and left Salvatore behind. Left my mother and all of us. A man can have a hundred motives, or no reason at all, to simply remove himself from the life he has made.

(p. 234)

Conclusion

As we have seen in the previous chapter, the *Literary Atlas* team identified key "plot-points" from our selected novels, and as we did so, recognized the possibility to further contextualize these in order to stimulate curiosity, creativity, and cultural engagement. This chapter has documented how the *Literary Atlas* project did so through the practice of deep mapping, a process that strongly complemented the approach of plotted reading, effectively putting the selected context back into the chosen texts through assembling multiple sources of data and juxtaposing them in a range of ways.

Unlike conventional deep mapping processes, this was a process of not simply mapping places, but mapping "plots". It was the composition of cartographies emergent through the coming together of the crafted words of an author, and the corresponding geographies of a reading team sensitive to the importance of geography to the dramatic arc of the fiction. In this way, this deep mapping of plotted reading can be seen to add new complexity to this cartographic composition, creating a form of "deep geography" (after Harris, 2015) which combines what the modern categorizes as "real" with more amodern "read-realities".

According to Harris, "deep geography" refers to the coming together of "deep mapping, spatial storytelling, and spatial narrative" (Harris, 2015:28). Combined with the insights of this chapter, we can suggest that such deep geography acknowledges that the world of which we are a part is "deep", a many (rather than a one) we are not "distanced" from, but co-ingredient with in a complex relationality which we can will never quite be able to fully comprehend. From this basis, doing "deep geography" can be understood as a *practice* of production (actively compiling mapping, storytelling, and narrative), but also the *output* of this practice – a deep(er) geography is the result of this ongoing composition (after Anderson, 2010, 2015, 2021) of mapping, stories, and curation for a specific location.

The visual representation of plotted readings can be seen to add new layers to the verticality of any deep geography – a cumulative assembly, and perhaps even subsequent co-constitution, of *spatial storytelling*, which we would understand to be the plotted reading of the crafted words of the author; *deep mapping* – or the compilation of a range of extra "data sets" to "identify, relate, and expand" these plots, and the curation of an explicit *spatial narrative* – or in this case plotline – to give a new order to these materials and enable moments of insight or illumination to new audiences. Recognizing this, the deep mapper of plotted readings comes to be further moved away from their role as distant producer of maps and their subsequent analysis, and towards a more entangled and embodied position as the curator of deep(er) geographies and their subsequent ambiguities – maps as processes and products which have the potential to communicate new knowledge to new audiences. As this new role emerged for the *Literary Atlas* team, we reflected further on our abilities and ambitions in this area – to what degree could and should deep mapping extend into more traditional artwork composition, and what could this potential offer to insights into literary geographies? Energized by this creative current, Chapter 7 explores how *Literary Atlas* expanded cartographies into commissions and collaboration.

Notes

1 As Pullman puts it, this borderland combines "our memories [and] the associations we have with this or that particular word or landscape" (2017:259).

2 As critic and author Jan Morris sums up, this view would hold that "authorship [i]s irrelevant to a book, that a work of art [stands] existentially on its own, liberated from its creator and unneedful of an audience" (2009:173).

3 Qualitative methods were pertinent to our needs as they enabled us to explore:

> individual's own accounts of their attitudes, motivations and behaviour, . . . the meanings and interpretations given to [their writing and] how these are put together, more or less coherently and consciously, into frameworks which make sense of their experiences.
>
> (Hakim, 1987:26)

4 In many senses, therefore, interviews are a variation of "normal" interactions between people, they are dialogues through which knowledge emerges, predominantly through verbal exchange. Interviews can range from a systematic interrogation akin to a cross-examination by authorities, through to informal "conversations with a purpose" (after Eyles, 1988).

5 Which highlighted the importance of top left impact. Western users traditionally start their reading here as they are influenced by the etiquette of more conventional texts, but as Laja (2022) suggests, then adopt an "F-style" reading, and require regular titles to command their attention.

6 As Anderson acknowledges, in some way, the idea of plotlines has resonance with the practice of "songlines", a reference to aboriginal ways of navigating territory, history, and community. It is important to note that despite the parallels, perhaps even direct resonance between plotted readings and indigenous ontologies (see, for example, Bawaka Country et al., 2015), there is "no precise analog in western culture" (Greenspan, 2011: 155) to songlines themselves. Much Western understanding of songlines derives from Chatwin's book of the same name (1987), which many scholars now actively seek to identify as a "rogue text" (e.g. Nicholls, 2019) constituting "an act of epistemic violence" (Nicholls, 2019:22) to the indigenous knowledge it purported to define. As Swift (2021:10) points out, it is notoriously difficult to fully engage with "First Nations' cultural practices from a Western standpoint". In her guarded interpretation of the songlines phenomena, she suggests the Australian continent "may be seen as a network of stories that are variously known as songlines or dreaming tracks . . . As well as recording the activities of the ancestors, the songlines function as mapping devices, recording locations of waterholes, mountains, hunting grounds – all that is needed to survive in the land" (Swift, 2021:9).

7 Expanding the field

Visualizing plots through cartographic art

As we have seen in the book thus far, currents of culture, space, cartography, and creativity have had a multiplicity of consequences for scholarly work: these currents have changed what academics study, how they study it, what methods they use to gain understanding, and what media they use to share it. These currents not only remind us that conventional writing and mapping remain both an important form of creativity and a vital means for method and communication (see Chapter 1) but as these currents take hold, they have also significantly expanded the means and media of academic expression.

As we have seen through the deep mapping process (as outlined in Chapter 6), through technological change (e.g. advances in digital platforms) and concomitant reorientation to the possibilities offered by existing media (e.g. the use of sound and art in many forms), new opportunities for intellectual mobilization and reflection have emerged. As this occurs, the contemporary academic role has expanded from a pure emphasis on rigour (premised on in-depth theoretical engagement, precision of argument, and detailed explanation), towards a wider acknowledgement of the power of situated knowledge, and even the capacity of ambiguity to inform and move an audience. Within these new circumstances, academics can now choose whether they wish to explain or enchant (or indeed a combination of the two); in their expanded role, they can continue to prioritize interpretation and analysis (after Hawkins, 2011), or "become creative practitioners in their own right" (Hawkins, 2011:464).

This expansion in academic choice opened up a range of innovative possibilities for the *Literary Atlas* project. As we continued the practice of deep mapping (see Chapter 6), the *Literary Atlas* team became enthusiastic about the potential of different forms of cartography (in the broadest sense) to understand our novels and their plot-points. For example, as we played with ideas from Trezza Azzopardi's *The Hiding Place* (2000), we began to consider how memory, place, and loss could be visualized; we wondered, for example, whether a three-dimensional deep map could be compiled, perhaps through the layering together of incomplete "jigsaw-maps" which might suggest the fragmentary nature of remembering and its influence on our person–place relations. In relation to Alys Conran's *Pigeon* (2016), we reflected on the title character's reconstruction of his own

DOI: 10.4324/9781003321248-7

identity through the piecing together of a stone wall; we contemplated how it might be possible to produce oversized "Lego-style" bricks which could be used by the public to construct "(insta)grammable" temporary walls (for example, at schools). These bricks could have synonyms for "wall" (e.g. "bridge", "connector", and "circuit") stickered to them in Cymraeg and English. In each reconstruction, the possibility could be offered of new ways to define the notion of what a wall is, or could be.[1] For other novels, we considered how it might be possible to repurpose seats from an old Wolseley Hornet, add a green screen, and enable users to pretend to drive Louis Knight's car over the hills from Aberystwyth to Ynylas (for Pryce's *Aberystwyth Mon Amour* (2001)); or how it might be possible to use drone footage to get a "Green Man" view of the wild hills of Ceredigion (to gain a mythological or otherwise non-human perspective on Griffith's *Sheepshagger* (2001)).

We were aware that these tentative forays into the "expanded field" of being a "creative practitioner" (or even more boldly put, of being an "artist") were exhilarating, but also (con)fusing. We became increasingly aware of the potential for exploring our skills, pushing our limitations, and closing down our gaps in knowledge in relation to artistic languages and craft, but also that this was an interdisciplinary excursion we had no existing collaborators to learn from, work with, or defer to.[2] We wanted to remedy this, and focus on the ways in which our (expanding) role could collaborate and converse with more established expertise. Reflecting on the project more broadly, we hoped that in expanding our skills we may also be able to expand our audience, stimulating interest in *Literary Atlas* beyond the academe. As Hawkins puts it, we saw the opportunity to develop "the long appreciated possibilities that creative practices offer for enrolling non-specialist audiences in geographical causes" (Hawkins, 2015:248).

To this end, we actively sought to establish relations with professional artists in order to explore new ways of creatively mapping the *Literary Atlas* resources to date. This chapter continues by outlining the initial process of commissioning artists, and the lessons learned from creative practices as method. The chapter introduces the reader to the artists, the novels they conducted plotted readings for, and the artworks emergent from these. Where artists have added expository context, these too are shared. Later in the book, Chapter 12 reflects on the process of displaying these artworks to new publics, and their responses to them; at this stage, however, the chapter concludes with reflection on the insights gained from collaborating with artists on the creative cartographies of our plotted reading borderlands.

Commissioning

To begin this process of artistic collaboration, a "Call to Artists" was established which outlined the *Literary Atlas'* plotted approach to literature. This call detailed how each commissioned artist would engage in their own plotted reading of a selected novel, draw on the deep mapping resources produced by the *Literary Atlas* project more broadly, and attempt to produce a creative cartography of the emergent borderlands. In more detail, artists were invited to

> visually convey a geographical aspect of their reading, broadly construed. Themes could include (non-exhaustive): novel location; novel theme; novel genre and atmosphere; character-place relations; relations between the "real" and "imagined" (or factual and fictional) geographies; colonization; language; movement; nationality; conflict; borders (etc.).
>
> (Call to Artists, Literary Atlas, 2019)

We outlined how, through this commissioning process, we wanted to playfully interpret traditional notions of cartographic mapping and explore the possibilities of visually communicating the relations between reader and plot. We outlined how we were open to more abstract or ambitious interpretations of the novels and/or their relationship with place, and were inspired in part by the work of Solnit (2010), DeGraff (2015), and Harmon (2003).

Following the placement of our call in a variety of artist forums, we received 62 responses, with some artists expressing general interest in contributing to any of our selected novels, and some honing in on specific "plots" (due to a personal connection with their place, theme, author, or history). We chose 11 artists, based in part on their ability to engage with different media (and thus hopefully produce a range of imaginative cartographies in different forms), but principally because particular artist statements and preceding work seemed to usefully expand or otherwise challenge our existing ideas. The twelfth collaboration was designed to be different, allowing non-experts (in this case, undergraduate students from a variety of disciplines) to experiment in this process of imaginative cartography. We hoped this collaboration would operate as a mini "pilot" study, developing insights and inspiration to run workshops with the public later in the project (see Chapter 12). Details of the twelfth collaboration will be introduced in detail below.

As we hoped this phase of the *Literary Atlas* would, to some extent, be a collaboration between commissioned artists and the project team, we initially set up a meeting to discuss novel selection, the deep mapping process to date, and our own thematic ideas that had emerged from this process. From this basis, the artists were given autonomy to interpret these materials into a work in whatever form, size, or medium, they felt inspired to produce. After a novel was chosen, the artist was given a copy of the book, access to the beta version of the *Literary Atlas* website, its synopses and plotlines (see Chapter 6), and the resources to visit any corresponding geographies they felt it important to attend. The commissioned artists, and their biographies, are initially outlined below, along with their artworks (which are also produced here www.literaryatlas.wales/en/art/). Although it is possible to engage with these artistic cartographies without explanation (in the enchanted geographies tradition, see Woodyer and Geoghegan, 2012), we nevertheless offered the artists the opportunity to explain or otherwise comment on their maps if they felt it appropriate. These responses are outlined along with each artwork below. Following this presentation of the artists, artworks, and expository comments, some insights from these collaborations will be offered.

Artist biographies and artworks

Richard Monahan "Et in Arcadia Ego"

Originally from Devon, Richard Monahan studied textiles at Camberwell College of Art, before moving to Swansea to study Fine Art in 2001. Since 2004, he has lived and worked as an artist in Swansea. His practice is primarily focused on narrative drawing with elements of painting and print contributing. Through this interest, he has exhibited both in the UK and abroad. Latterly, his practice has widened to encompass drawing research and teaching.

Richard Monahan "Et in Arcadia Ego"
2 × 60cm × 85cm
Pen, pencil, masking fluid, card, rain, metal leaf on card
A response to Malcolm Pryce's *Aberystwyth Mon Amour*
Artist statement (extract)

Reading can be a very subjective experience. Each word in a sentence cannot convey precisely the same meaning to each reader, but is distorted sometimes by personal experience. As a reader, this means that you might emphasize one aspect of a text to a disproportionate degree of importance, so that it towers over and distorts/subverts meaning. The re-creation of text in this way results in the process of reading being as much a portrait of the reader as it is of the author. The missing link in communication, the gap between intention on the part of the writer and reception

Figure 7.1 Richard Monahan "Et in Arcadia . . .".

on the part of the reader, offers nuance to the reading experience, allowing the world created by the author to be fully inhabited and accepted by the reader. It is within this space that I have created this map.

When I made the trip to Aberystwyth, the experience of water was intensified. The steep beach results in waves that seem to crash almost into the town itself, eating it up fraction by fraction, and it rained fairly constantly. This defiance of or submission to geography, the role of the elements in the book and the situation of Aberystwyth on the edge of a land, all combine to give the reading experience a sense that we look down, God-like, on the action as it is played out. With an air of benign but futile hope, we see that the players in this game are each ultimately and fatalistically wrapped up in their world, wrapped up in a sequence of events that are doomed in the face of the greater turning of the earth and movements of space. The smallness of human endeavour and the underlying existentialist tone that pervades the book drew me to consider Sospan, the ice cream seller, as central to understanding the story. His motto, *Et in Arcadia Ego*, positions Sospan as a part in this first story, and in this sense a friend to the reader. I have placed him looking out to sea.

If the sense of looking down on the story established a schema for the map philosophically, it also provided the structure for the drawing technically. The use of parallel perspective enables a positioning of the viewer over the action, but close to it also. So the town is brought towards us, before a reversion to a more conventional perspective for the background. The use of drawing as a medium was also useful as it displays the history of the process; displays the trying through time that

Figure 7.2 Richard Monahan ". . . Ego".

is core to what creativity is. In this sense, it was important that the map was a map not only of the story but of the handmade making of the map as well. I wanted the map to be demonstrably a map of physically and mentally figuring something out, where flaws are not effaced, and where process vies with illusion. This approach allowed me a freedom of expression to create a map that was not entirely readable in the sense of navigation, the reason for a conventional map. This map constantly revisits its own process and its own sense of being an object. In doing so, it is able to create a tension between the book locations and the subjective experience of reconstructing an idea from text, through a process of interpretation, and through to my consequent attempt to re-communicate via marks on a piece of paper.

Rhian Thomas "Measuring the Distance"

Rhian is an architect and co-founder of ALT Architecture, a studio set up to explore the "edge" of architecture. It is within this fertile ground between architecture and

Figure 7.3 Rhian Thomas "Measuring the Distance".

other disciplines that is the most interesting, as it provides opportunities for exploration, innovation, and creating new dialogues between art, landscape, and theory.

Rhian Thomas "Measuring the Distance"
45cm × 70cm
Mixed media
A response to Raymond Williams' *Border Country*
Artist statement (extract)

Measuring the Distance was made following a number of psychogeographical excursions to "Border Country": the landscape was walked and recorded to enable perception and an intuitive response to sense-making to inform ideas on how to visualize the places held in the novel *Border Country* by Raymond Williams, and convey the power of a human attachment to place.

Historic maps were torn and layered to produce an abstracted profile of the landscape in plan and elevation, depicting the geographical character and location of the novel. The layering is symbolic of the geological composition of this wild and remote terrain but is also evocative of a land that bears the imprint of human occupation over generations. There are few places in Wales where this relationship between mankind and the land is more visually evident than the Black Mountains, and this is described by Raymond Williams in *People of the Black Mountains*:

> See this layered sandstone, in the soft mountain grass. Place your right hand on it, palm downward . . . with this red earth, this place was received and made and remade. Its generations are distinct but all suddenly present.

The work also seeks to address the two counterpointed timelines obvious in the novel; that of Will and his parents Harry and Ellen, versus that of Matthew (Will) returning as an adult to visit his ailing father. Notions of local knowledge, distance, measure, and time were visualized in the map by equating measure and distance to footsteps in defining the map's scale. This is representative of how, through the text and experiencing this landscape, we get a real sense of how physical distance is measured relative to key markers in the landscape, and how the characters living in the valley have an inherent understanding and connection to place. It is a place where time is understood and measured by the ritual of trains passing through the valley, the changing seasons, and the agricultural calendar.

The trainline in the map symbolizes a second scale of time and distance evident in the novel, portrayed through Matthew's return and departure to the valley by train from London. He struggles to distinguish between this landscape as a romantic image in memory and the reality of this landscape he knows so well, an ever-changing entity influenced by shifting cultural and temporal dimensions, and he is concerned about how to "measure" change. This dual meaning of "distance" and "measure" in the novel, and the notions of time, space, and temporality which inflect every page, inspired the title of the art map, along with the closing statement

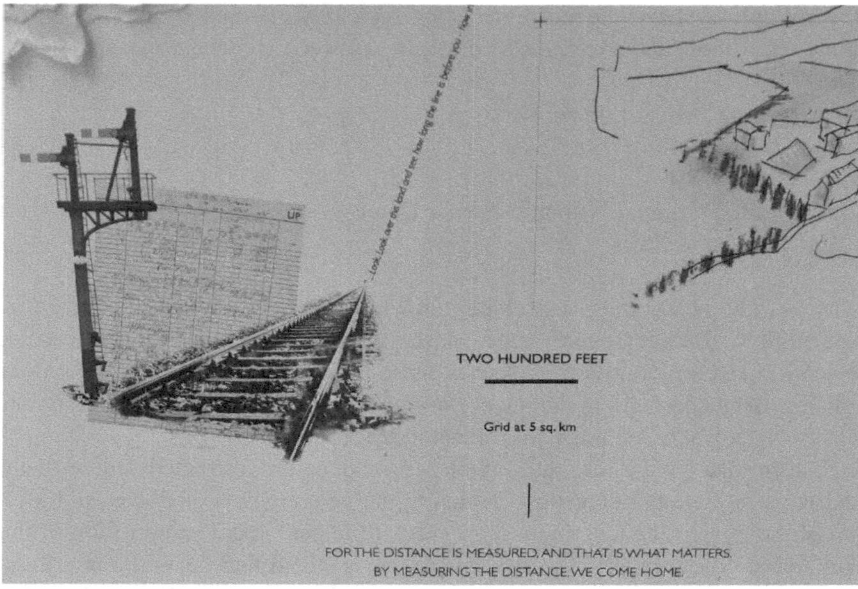

Figure 7.4 "Measuring the Distance" (close up).

in the novel; "for the distance is measured, and that is what matters. By measuring the distance we come home".

George Sfougaras "A Personal Map of *The Hiding Place*"

George is a painter and printmaker. His work explores issues of human migration, change, and cultural inheritance through representational art which draws heavily on cultural symbols and historical sources. He co-ordinates *Focus on Identity International*, a group of 19 artists from 12 countries.

"A Personal Map of *The Hiding Place*"
120cm × 100cm
Archival ink and acrylic on Fabriano paper
A response to Trezza Azzopardi's *The Hiding Place* Artist statement (extract)

Trezza Azzopardi's father was from Malta and my family are Greek. This duality creates fertile ground for asking interesting questions, but not necessarily finding answers. There is a tension between too many allegiances in belonging to two cultures and to none. I hungrily read the book and became submerged in the emotional landscape of the novel. The setting had a very distinct physical aspect. Tiger Bay, Butetown, the slums of the old Cardiff Docks. Old Victorian houses in various stages of decay provided the backdrop for a human drama that one could not but feel drawn into. The smells of the surroundings, the dull, joyless houses, purely

Figure 7.5 George Sfougaras "A Personal Map of *The Hiding Place*" (without lightbox).

functional, and then only just so; the logistics of keeping a family, when the main earner is a gambling, immoral, and self-absorbed loser; the innocent descriptions of the horrors of paternal neglect and violence.

I started to draw simple descriptive maps. Their emergence and development were entirely unexpected. I can only suggest that they came from a deep need to revisit aspects of my past and to understand our family history. My mind turned to places and experiences that I have held onto, or more accurately filed for future

Figure 7.6 Developing "A Personal Map of *The Hiding Place*".

reference. Echoes of my childhood and places half-remembered emerged in a simplified linear form, sometimes superimposed over archetypal faces of my younger self and my parents, or in the form of illustrations of places half-forgotten, lost in time, or destroyed by war.

Undertaking this project has strengthened my belief that we need to look deeper into what constitutes a good life and what aspects of our lived environments testify to our values. The old communities of Tiger Bay, the inspiration for this wonderful novel speak volumes for what happens to us as people when we are without direction and when the moral basis of our existence is shaken by poverty and violence.

Seán Vicary "Sitelines"

Seán Vicary's work deals with landscape, place, and our interaction with the "natural" world. He often explores this relationship through found objects and fragments of detritus, manipulated in a virtual space to create animated assemblages. These act as triggers for the viewer, sometimes suggestive of a wider narrative or the hidden processes at play behind the visible. Vicary is the recipient of a 2017 Creative Wales Award from the Arts Council of Wales and is researching contemporary manifestations of Genius Loci and how they might challenge our human-centric

Figure 7.7 Seán Vicary "Sitelines", see www.literaryatlas.wales/en/art/artwork/the-owl-service/.

viewpoint. Sitelines won the Gold Medal for Fine Art at the 2022 Eisteddfod (see Eisteddfod, 2022).

Seán Vicary "Sitelines"
HD video & animation
Music: Ceri Owen-Jones, Elsa Davies, Ceri Rhys Matthews
A response to Alan Garner's *The Owl Service*
Artist statement

I grew up with Garner's early novels. Looking back, I can see how they suggested a framework for mapping my own childhood boundaries: defined by ditch, field, and hedgerow. But the Owl Service was something else. Initially slightly too young for the book, my introduction in the 1970s was via the TV adaptation. I'm not sure I ever recovered. This work explores the novel's idea of the valley as a charged "reservoir" for unseen forces but also as a driver for Garner's (and consequentially my own) creative process; the immersion of oneself in a landscape/place providing a conduit and the act of creation a channelling of those forces. When filming, I was frequently aware of the sight lines between locations, recalling Roger's experiments in photography and his lining up of the hole in the Stone of Gronw with the far hilltop; this line of sight enabled a direct path between transmitter and receiver, providing an eerie optic onto past events. I've collected, photographed, and animated objects from the valley, including horsehair and lime plaster, handmade nails, and a witch bottle. Together these create a "receiver" for tuning into residual wavelengths and decoding any hauntological transmissions.

Amy Sterly "The Missing Piece/Y Darn Sydd ar Goll"

Amy Sterly is a printmaker and sculptor originally from Chicago, now living in a small community in Wales. She tries to subvert the idea of the rural idyll through her prints and her recent sculptures have reinvented the nature of the book into something tactile that triggers memory and emotion.

Amy Sterly "The Missing Piece/Y Darn Sydd ar Goll"
200cm × 180cm × 66cm
Dictionaries, Hans Christian Anderson storybook, paper, cassette tape, slate, matches, comics, hanger, embroidery thread
A response to Alys Conran's *Pigeon*
Artist statement (extract)

Being involved in the *Literary Atlas* project has been an exciting and inspiring journey that has helped me create a new direction for my work and it also helped me explore new connections and meanings between literature, art, and the concept of mapping. The interdisciplinary nature of the project forces one to think about the nature of fiction and its place in the real world and how it connects with the way we imagine the story we are reading. *Pigeon* connected to all the themes and ideas I have been working on recently and brought me to the mountains of North Wales with new eyes. The landscape and characters came alive and the words and images flew around my head and an idea of a new artwork formed. Certain moments were amplified into the symbols and themes of the book, and Welsh and English words

Figure 7.8 Amy Sterly "The Missing Piece/Y Darn Sydd ar Goll".

were woven into a single tapestry. The old Welsh samplers came to mind and the dry stone walls that cover the hillsides. The weight of the words was like stones and built a foundation for the novel. The characters "taste" the words and spit them out again to create the narrative. So many ideas, a nest of ideas, to make an artwork, out of books, out of scraps, out of keepsakes, and out of paper aeroplanes cut from Welsh books. Built firmly on the ground and hanging by a thread.

Valerie Coffin Price "A Rioter's Walk"

Valerie Coffin Price is an artist-letterer based in Cardiff, Wales. Her work responds to the poetic resonance of language and its connection to the environment and cultural identity. For the past 20 years, she has been working with poetry and literature as inspiration, source material, and content for her art practice.

Valerie Coffin Price
"A Rioter's Walk"
Sketchbook
A response to Amy Dillwyn's *The Rebecca Rioter*
Artist statement (extract)

Collaboration and cross-arts working are essential elements of Valerie's art practice, experimenting with, and reflecting on, different ways of seeing which create a conversation between the environment, people, and her response to a specific place. The folding sketchbook has been where much of Valerie's work has taken shape on

Figure 7.9 Valerie Coffin Price "A Rioter's Walk".

recent projects. The twists and turns of the escape journey in The Rebecca Rioter, linking and separating places, changing and enduring through time, are mirrored in the interplay of art and text in the Rioter's sketchbook, both being reflected in the structure of the folding book itself. Her approach on this project involved a series of walks and investigations, approached through a process of being in conversation with the book and its writer, the journey and its sites, in effect layering place and imagination, language and image. This generated a series of experiments from drawing and photography to film using a GoPro and mobile phone as well as the folding sketchbooks.

John Abell "Hiraeth for Beginners"

John Abell is a visual artist most identifiable for his monochromatic prints and vivid paintings exploring life, love, lust, and the human condition. Abell was born in Cardiff in 1986 and graduated from Camberwell College of Art and Design

Figure 7.10 John Abell "Hiraeth for Beginners".

in 2009. He has exhibited widely and his prints and publications are included in museums, libraries, and national collections worldwide.

John Abell
"Hiraeth for Beginners"
244cm × 122cm
Woodcut, acrylic, ink, and watercolour on paper
A response to Tristan Hughes' *Revenant*
Artist statement

I didn't want to focus on traditional cartography. Instead, I decided to visit Penmon, where the novel takes place, and charted major plot-points leading to the denouement, which occurs on Penmon beach and made visual interpretations of them with a focus on creating an image that works outside of the context of the novel, but easily understandable to someone familiar with the text.

Locus "The Blue Hawk"

Locus is the combination of composer, musician, producer, and visual artist Angharad Van Rijswijk (formerly Trwbador, currently Accü), and composer and musician Richard James (formerly Gorky's Zygotic Mynci). They formed in August 2015, and have been working together constantly since then, mainly on radio, live, and exhibition installation projects. Their first album was released in January 2019.

Locus "The Blue Hawk"
Audiovisual work
18cm × 29cm × 20.99cm digital prints with 17min soundtrack
A response to Niall Griffiths' *Sheepshagger*
Artist statement

We (Locus) have been working directly with concepts of landscape, memory of landscape, space, and psychogeographical elements ever since we've been working together, so when this opportunity arose we were excited about being a part of such an interesting project, with regard to the *Literary Atlas of Wales*. I had read Niall Griffiths' *Sheepshagger* some years previously and very much enjoyed it, and as we had recently relocated to West Wales, we thought an audiovisual response to this book would make sense. We visited the obvious locations mentioned in the book, but also with the insight provided by Niall and the *Literary Atlas of Wales* team, the approximate, half-imagined conflations of linear and creative landscapes which serve the narrative and the prose. From field recordings and photographs we took on the trips, we then came back to our studio to compose and assemble a series of sound works, and Angharad created the visual artworks from the photographs, which were also filtered through our own memory and instinctive emotional response from our immersion in the rich and vivid prose of the book and the landscapes we visited.

Figures 7.11 and 7.12 "The Blue Hawk" (selected images, see www.literaryatlas.wales/en/art/artwork/sheepshagger/).

Liz Lake "Endnote"

Liz Lake is an artist who works across printmaking, sculpture, and installation. Her work often looks into material processes, and responds to certain landscapes and places. Recent projects include being shortlisted for the Ironstone Art Prize, a

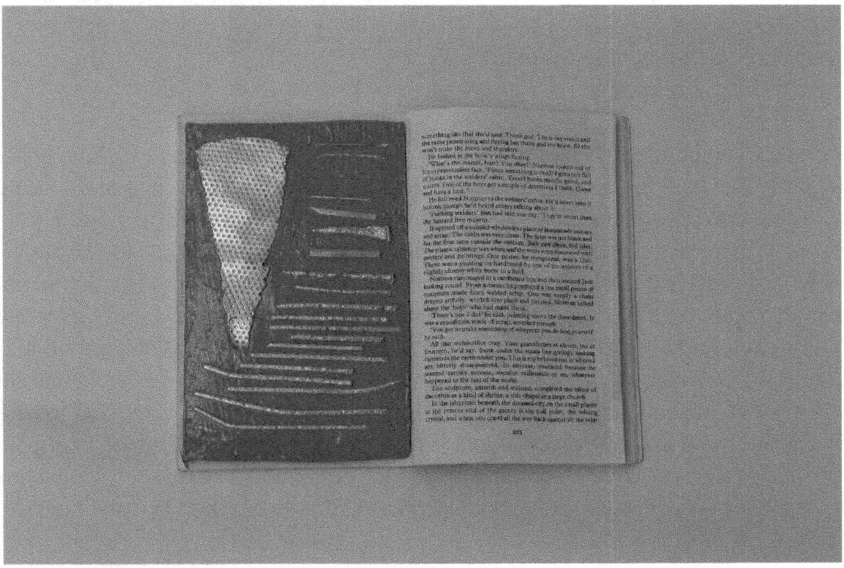

Figures 7.13 and *7.14* Liz Lake "Endnote".

24-hour residency at RRU in Liverpool and showing work as part of Whitstable Biennale Fringe.

Liz Lake
"Endnote"
Cast bronze, tin plate, enamel paint, steel, screen-printed and photocopied archive documents
A response to Christopher Meredith's *Shifts*
Artist statement (extract)

Endnote is the piece I have made in response to the novel *Shifts* by Christopher Meredith. The book is set in 1977 in the steel mill Ebbw Vale and documents the end of steel production and the move to the newer process of electrolytic tin-plating. I was interested in finding out more about the setting of the book and visited the Ebbw Vale Works Museum Archives to look at documents and images relating to this period. I was also drawn to a section of the book that describes the "welders' cabin", decorated with pieces of artwork they had made from scrap metal. For me, this passage highlighted the crossover where skills used in commercial workplaces meet craftsmanship and aesthetic judgement, and questions when and how we place value and merit in objects and materials. My piece combines tin-plated cast bronze with enamel paint, steel, and sections of archival documents printed onto tinplate and paper. With thanks to Ebbw Vale Works Museum Archives for permission to use their text and images.

Iwan Bala "Enlli: A Learning Aid"

Iwan Bala is an established artist, writer, and lecturer. He has held solo exhibitions annually since 1990, participated in many group exhibitions in Wales and abroad and is represented in public and private collections. His work was exhibited in four Chinese cities in 2009. He has published books and essays on contemporary art in Wales and is a frequent lecturer on the subject. He has often presented and been interviewed for television. Iwan Bala is cited in most published compilations on contemporary art in Wales.

Iwan Bala
"Enlli: A Learning Aid"
120cm × 90cm
Mixed media, ink, pastel, and collage on Indian Khadi paper
A response to Fflur Dafydd's *Twenty Thousand Saints* and Ynys Enlli (Bardsey Island)
Artist statement.
Enlli: A Learning Aide.

"You who are in the traffic of the world: can you guess
the thoughts of an islander".
Brenda Chamberlain.

Figure 7.15 "Enlli: A Learning Aid".

These are words expressed in Brenda Chamberlain's book "Tide-Race" (1960) which is a diary of her time on Bardsey Island from 1947 to 1960, and my first edition copy includes her own illustrations. She was an admirable artist and poet, twice winner of the Gold Medal for Fine Art at the National Eisteddfod in 1951 and 1953, and in my work for this project, I have collaged into the image, copies of her drawings which I sought to emulate. I realize that my task was to read and comment visually on Fflur Dafydd's novel "Twenty Thousand Saints" of 2008. It seemed to me that the novel is a modern fiction that (whilst being a great televisual or filmic opportunity) was about those who came to the island rather than the island (or its people) itself, the "Enlli Profundo". Brenda Chamberlain immediately came to mind as someone, an artist and poet, who had decided to live on the island while it was still very much a place "other" than the mainland. The fascination that this island holds, is well expressed in the novel by Fflur Dafydd, yet I somehow feel that it reflects contemporary issues transplanted to Enlli, rather than the involvement of time, history, and memory, which, though touched upon, are not fully realized. This might be deliberate, to show Enlli as a microcosmic Wales, where local culture is subsumed for the ascendancy of a hegemony from beyond its frontiers. Fflur Dafydd is concerned with "the traffic of the world" that Brenda Chamberlain abjured during her residence on Enlli. I understand how the island "stands in" for Wales in the political discussion, my work over decades has suggested a similar view of Wales as an island always on the horizon. An ideal state that is permanently at a distance, unreachable. The island of Gwales (Grassholm) in its usage *in Y Mabinogion* has been a recurring theme. Much of the symbolism of these earlier works are reintroduced in "Enlli; A Learning Aid", the island at a distance, the stylized black boat/cauldron, and the ladder.

Despite the distance of time (and indeed, of political agendas) between Brenda Chamberlain and Fflur Dafydd, there are themes that reoccur, in particular, the excavation of hidden truths or of lies. As a place of mythical memory, Enlli could not fail but be the inspiration of uncoverings. The very fact that it is still a difficult place to get to (I tried twice in 2017, but the hidden tides were against me) makes that an enduring fact. However, to compensate for my lack of lived experience, I approached a trusted friend, the poet, and writer Sian Northey, who recently completed a writer's residency on the island for some words that I could also add to the "map" or "learning aid" that I was trying to construct. She generously sent me all she had written, as yet unpublished. From a poem called "Rhwng Dau Gae yn Enlli", (Between Two Fields in Bardsey), I chose the closing lines:

Yma, lle mae esgyrn yn y pridd, mae'n hawdd
mynd o un cae
i'r cae nesaf.

My literal translation would be

Here, where there are bones in the earth, it is easy
to go from one field
to the next field

These deceptively simple lines encapsulate for me the feeling I wanted to bring forth in the artwork, they refer to the presence of the dead in the continuing existence of the living. How the world we inhabit can easily slip into the time of before and vice versa. It is a theme that has dominated my practice for several decades, one I am also researching now for a doctoral thesis. My concern with what I have defined as "custodial aesthetics" is the use of contemporary practice to delve into a past (a forgotten, yet real, history) that might inform the future. I feel this a necessity, in particular for a minority culture, whose history is ignored or derided. My

Figure 7.16 Joni Smith "Concrete Ribbon Road".

work has propagated the idea that Wales is itself an island, with much of its history buried or submerged, Enlli, is therefore, not only a microcosm of cultural memory in itself but also a metaphor for "All that is Wales".

In the image, I have sought to create a revealing of the layers of history that echo the layers revealed by archaeologists, a theme in the Fflur Dafydd novel, that I assume to be an unearthing of the history of Wales that still lies buried, whilst also being an examination of post-Devolution Wales. Within the layers, the water's current, the shifting earth, there is movement. But there is a sense of an almost unbridgeable gap between two worlds, linked only by words of poetry, and the fragile ladder that writing provides. The ladder in the image leads to the boat/cauldron of rebirth, that might take us to the promised land.

Joni Smith "Concrete Ribbon Road"

Joni Smith is a contemporary artist known for her paper and installation works that explore and rework the map. Originally from Leamington Spa, raised in the hills of North Wales and currently residing in Norwich – Joni's fascination with a sense of place threads throughout her work.

Joni Smith
"Concrete Ribbon Road"
80cm x 60cm
Collage using maps and atlases

A response to Lloyd Jones' *Mr Vogel*
Artist statement (extract)

The piece is based on lots of different aspects of the book, whilst looking chaotic, every cutout refers to an aspect of the book. Anglesey is mentioned frequently in the story and in the piece it appears squeezed between the horizontally laid out land of Wales (which has been morphed into a mountain with a winding road climbing up and up. The land parts are cut out of the map of Anglesey and in their place are "gems" – fragments of land – castles/sheep/lighthouses . . . all parts of the story stored like nuggets neatly in the crevices and nooks of the winding roads of Anglesey. The title "Concrete Ribbon Road" is a direct quote from the book (p. 13). It is the part of the piece that ties all of the elements together – a pathway that you can view the work from. It traverses up the mountainous terrain (of the up-ended Wales). It weaves through the sky chart and ends up as a chaotic ball of floating twine. I like the idea that unless you have read the book you cannot decode the meaning – To quote Henry Adams, "Chaos was the law of nature, order was the dream of man." This is how I read the book and how I wanted to depict the piece – a kind of making sense out of a chaotic world and sometimes revelling in that chaos.

The Twelfth Collaboration

For the twelfth collaboration, the *Literary Atlas* team wanted to explore how art could be a "dialogue" and a "doing" (after Hawkins, 2011) for those without formal artistic training, in particular for undergraduate students from different disciplinary backgrounds. To that end, we recruited students from a range of subject areas adjacent to and overlapping with literary geographies (history, human geography, and English literature) to explore how their disciplinary suppositions may influence what they considered literary geography (and mapping in particular) to be, and how these different insights may combine to produce a particular cartography. These four students (employed through the Cardiff Undergraduate Research Opportunities Programme): Marcus Sams (English Literature), Holly Cooper (History), Hana Undy (Human Geography), and Alice Savage (Human Geography), collaborated for the piece "Motive and Opportunity", based on Gallie's *Strike for a Kingdom* (1959).

Cardiff University Student Project
"Motive and Opportunity"
200cm × 200 cm
Mixed Media Crimeboard
A response to Menna Gallie's *Strike for a Kingdom*
Artist statement

Our artwork is based on *Strike for a Kingdom* by Menna Gallie, a murder-mystery novel set in Cilhendre, a fictionalized version of Ystradgynlais, during the 1926 General Strike. To represent the contextual layerings of the novel, we have chosen to physically layer the historical, fictional, and geographic context upon one

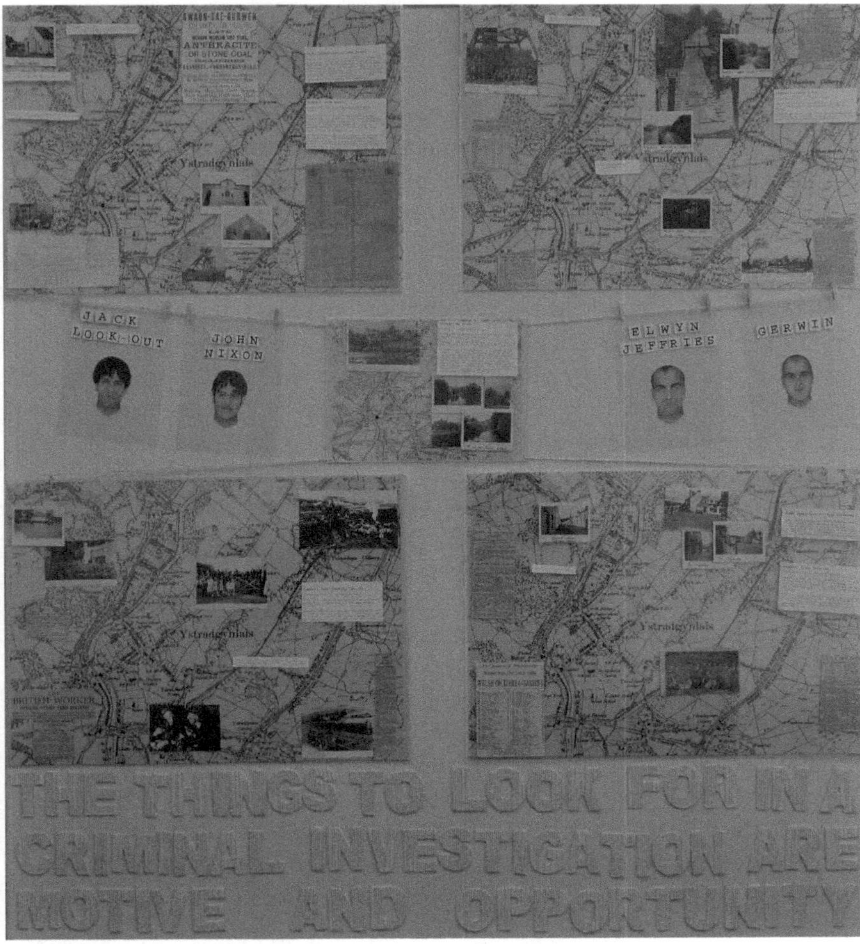

Figure 7.17 Cooper, Sams, Savage, and Undy "Motive and Opportunity".

another using acetates. We have arranged these layers into the novel's four main suspects. The top layers provide a profile of each suspect, including their motives, caricatures, and location to plot the events of Gallie's novel. The second layer groups both contemporary and historical images of Ystradgynlais to ground the novel within the geographical locations in which it is set. The bottom layer introduces the historical information from an array of national newspapers and their reports of the strike. This contextual layering forms a deep map that enables the viewer to look at the information within its spatial context and view them all collectively, as Gallie presents them. Our art aims to provoke discussion between different perspectives rather than presenting a single statement of paper.

With respect to the 12th commission, we asked the students involved to explicitly reflect on their experience of interdisciplinary engagement and how their particular predispositions and skills were built on or challenged through the process of imaginative mapping. There follows an excerpt from this reflection:

> We curated our assembled materials on a board (including photos sourced online, resources gathered from our day visiting archives, and photos, maps, leaflets (etc) from our visit to Ystradgynlais). This board's scale lends the art piece a sense of significance and the cork material echoes the everyday materials often used to construct murder boards in film and television.
>
> Police advisors on set for TV dramas point out that such murder boards are inaccurate, and often dramatic, conceits. Despite this, shows about modern policing persist in presenting information in this stylised, artistic, and public manner. In television they can function as a visual method of expressing the genius of detectives. How characters interact with the "murder board" often serves a narrative function such as the color-coded displays represented in various episodes of TV drama "Homeland" which are assembled in a frenzy by character Carrie in sequences featuring her bi-polar illness (which acts to communicate the co-morbidity of creativity, brilliance and mania). This raises the question about how we want people to engage with our murder board? Ideally, we would like viewers to engage in the task of solving the mystery, as active participants. By providing various suspects we hope viewers will feel suspicion and consider who is responsible . . . . We decided to try and highlight the tension in the text between the stated methodology of criminal investigation, "The first things to look for in a criminal investigation are motive and opportunity", and the real "evidence" and bias the inspector uses to judge guilt, namely, class position. This forms the bias which places John Nixon "a good, superior boy" above suspicion. Menna highlights this tension herself by showing the investigator's class bias and bigotry before revealing that the suspicion cast on all the colliers was entirely misplaced as the real motive for the murder of Mr Nixon was unrelated to the class tensions derived from the strike.[3]
>
> Today the acetates we ordered were delivered . . . .We have practiced arranging the resources into multiple configurations:
>
> Plot and characters:
>
> a. character fact files (produced to be reminiscent of official police documents with character summaries noting motive and opportunity)
> b. the location of the suspect that night
> c. the path the suspect took as they moved during the night
> d. quotes from the text
>
> 2. Places
>
>   a. images of significant locations in the book (historical, printed black and white and contemporary, taken on polaroids)
>   b. meaning of places – polaroids have the significance or name of place written on them (for example "where the baby's coffin was buried")
>   c. lines connecting these images to the location at which they were taken

3. class tensions/historical context:

    a. newspaper clippings from British Worker and the British Gazette, material from archives.

4. Location

    a. This layer is not acetate, it is the A1 historic map which underlays the acetate layers.
    b. This layer viewed in isolation is plain, however we drew lines from the articles on other layers to the points on the map they refer to. This means when the layers are stacked these lines demonstrate the interaction between location and all subsequent information.

How we ordered these layers is not insignificant, through experimenting with different arrangements we found that the meaning derived from the art piece changed significantly if the order of the layers or what was on each layer was adjusted. A clear example of this is what the project could look like if, instead of sorting the acetates into four separate character narratives, we piled them all together, or viewed all of the plot and character layers together in isolation from other information. The interaction between layers is highlighted when layers are viewed in isolation as they appear so incomplete. However just by adding one more layer to the above image it begins to appear more fleshed out and could be viewed as an object of interest in its own right.

In the final piece selecting the order of the layers was a straightforward decision. As a group we decided that the plot and characters which the novel

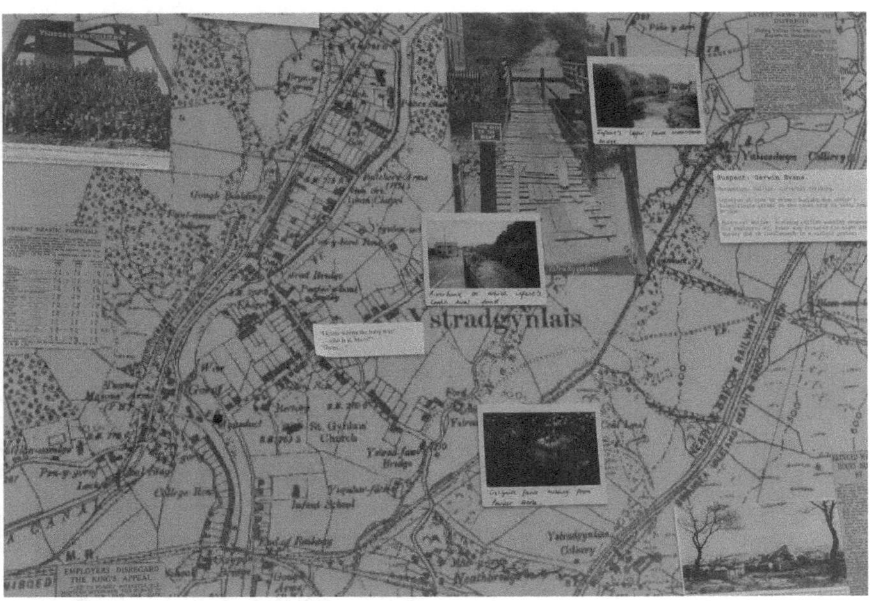

Figure 7.18 Layering "Motive and Opportunity".

> foregrounds should be the first layer, in order to make it the most visible. The location (the map) was the background to the novel and the historical social tension also underlay the novel's plot and so it seemed apparent that while essential to reading the art piece as a whole this was best displayed below the other layers (Savage et al., unpublished).

Commentary: creative cartography as method and output

The production of creative cartographies through collaboration with professional artists and others can be seen as an extension of the plotted approach to the *Literary Atlas*'s selected novels. Artists undertook their own plotted reading of a chosen novel, reflected on their own corresponding geographies, and drew on the materials compiled by the *Literary Atlas* team, to make further creative contributions to the conversations about the borderlands of English-language novels set in Wales.

Following insights from Hawkins (2011, 2015, 2021), it is clear that the production of creative cartographies makes explicit an aspect of artistic endeavour that is directly shared with other scholarly work; that is, artwork is not simply an output (i.e. the presentation of a finished knowledge product), but also a method (i.e. the means through which to create knowledge). Just as many academic students will have experienced with respect to the act of scholarly writing, creativity is not the passive writing-up of some "thing" that is already discovered, fixed, and known, it is an exercise in actively working with our chosen media (in this case, words, concepts, and empirical materials) to think through, affectively respond, and try out combinations of communication; it is through this emergent process that knowledge in some form or another is forged. Similarly, what is clear from the expository comments of the *Literary Atlas* collaborators is that the act of producing artistic maps is not a process of delivering an already considered and complete idea, but it is a process of active engagement with the media chosen (e.g. words, maps, sound, metal, paint, and canvas) through which lines and shapes form, ideas emerge, and (tentative) outcomes are paused upon. In this sense, the act of creating (be it writing, painting, playing music, etc.) is a practical method that can be used to form understanding; such a conclusion is acknowledged by Hawkins, when she notes with respect to Picasso that he categorized all his:

> paintings [as] but research and experiment. I never do a painting as a work of art. All of them are researches.
>
> (2021:11)

The process of creating is therefore significant beyond and before the importance of any artistic "entity/process" that emerges from it. It is the process of emergence that produces understanding and insight for the artist, and this acknowledgement is something that Monahan (earlier) wished to weave into "Et in Arcadia Ego", "demonstrably" making clear to his audience that the work emerged through "physically and mentally figuring something out" (see earlier).

Monahan's remark does more, however, than point towards the ability of the artist to draw attention to their commissions as the "art" of "work"ing (out), but also the artist's more general wish for their work to draw attention in the first place; in other words, to engage an audience *beyond* themselves. As Chapter 1 outlined, all culture realizes a vital part of its potential through engaging with an audience, thus beyond its significance in producing meaning for the artist, it also needs to command the interest of others. Each of our commissioned artists, therefore, had to "figure out" (after Monahan, earlier) what they wished to present to their potential audience, and what possibilities their chosen media offered for audiences to "read" the materials in various ways. In some sense, this engagement is practical. Unlike literature that is read (in the Western tradition) from the top left of a page and on horizonal lines, not all artwork has such often taken-for-granted etiquettes of engagement. In conversation, the artist of "Hiraeth for Beginners", John Abell, suggested how he liked his artworks to share the characteristics of traditional (paper) maps, offering the user no definitive point of entry. Prolonged "reading" of his work, he hoped, would prompt (perhaps clockwise) spirals of engagement (which may in turn offer particular connections to be made between aspects of his interpretation of *The Revenant*'s plot-points, see Figure 7.10, earlier). Beyond the practical, the students involved in "Motive and Opportunity" also reflected on how working beyond words made their choices for audience engagement more explicit; they wrote:

> A core question with our murder board was how to make a compelling image that reflects the complexity of the novel's slow unravelling who-dun-it plot while also being comprehensible to a viewer who hasn't read the book. Beyond concerns about legibility of the board, an ongoing concern was how we want this art piece to be read.
>
> (Savage et al., unpublished)

With this reflection in mind, what became fundamental to all the *Literary Atlas* artworks (and perhaps all forms of creative production, including academic monographs, presentations, and papers) is how they command attention from a potential audience, and related to this, what they want their audience to do following their encounter with it? It is as if artists, perhaps even tacitly, put themselves in the position of their audience, and ask: what am I being presented with – what is this work, and what work is it wanting me to do? Is it a question (for me to answer), a puzzle (for me to solve), or an answer (to a question I may not have even asked)? In short, what is my role here? In order to engage with the risk of their chosen form becoming "chaotic" and "cognitively overwhelming" (after Eanes et al., 2020:8–9) (a risk shared by deep mapping, see Chapter 6), these artists had to reflect on these questions, and take a particular position, as the artists producing "Motive and Opportunity" state:

> By virtue of the packed and chaotic collage aesthetic of a murder board, even when corralled into an ordered structure it is difficult to read – the path the viewer's eye should track is not clear. To a degree this reflects the real process in investigating a crime, the inspector in the novel is faced with a wide variety of information of varying relevance and accuracy. A weakness of the

> art piece is, therefore, that it fails to communicate the existence of parallel possibilities on the night of the crime, and perhaps that the most important information is lost and overwhelmed in the mess of context. But perhaps, in a way, that becomes the point.
>
> (Savage et al., unpublished)

When considering what position to take on the issue of user engagement, it is clear that artists can leave their works to "speak for themselves", offering little explanation or exposition for their potential audience (and some artists, like John Abell and Locus (earlier), are content with the artwork itself to be the key means of audience engagement in whatever way they feel appropriate). Yet it is also possible, as in the case of the student work and others such as Sfougaras and Bala, that extended expository notes can provide new layers and context (or "depth" perhaps, in the spirit of Chapter 6), in order to offer a "personal" instruction, explanation, and interpretation of the art. It is important to note, however, that reading creative cartography, like all map reading, does not conclude with the discovery of a hidden truth, but the emergence of a provisional one; the nature of this "take away" is predicated on how the chosen "lines" of the artist (be they in two or three dimensions, in sound, film, moving, still, and various juxtapositions of all) "touch the life" (after Bradbury's Dr Faber, 2008:108) of the audience member.

Thus, complementing the process of deep mapping (see Chapter 6), these creative cartographies, emergent from a plotted reading approach, do not produce in themselves, or seek to generate in their audiences, a definitive understanding of the borderlands, but rather contribute to an ongoing "conversation" about it (after Bodenhamer et al., 2015:21). Indeed, these artworks in themselves can be considered as the "new data, new perspectives, and new insights" (Bodenhamer et al., 2015:4) which can act to re-(and de-)compose the plot-points and plotlines composed by the *Literary Atlas* project more broadly. As Chapters 12 and 13 will reflect on in more depth, such repositioning of what knowledge is and can be – from simple chunks of insight that can be extracted from the world, or as fragments of illumination or enchantment that kaleidoscopically come together through our situation within it – is part of the vital significance generated through academics commissioning, and collaborating with, artistic practitioners.

As we have seen, the chapters in the first half of this book have documented *Literary Atlas*' moves through the various manifestations of the plotted approach: from the production of scholarly analysis, through synthesis, and into increasing ambiguity. Mobilized by the cultural, spatial, cartographic, and creative currents, the *Literary Atlas* has embraced plurality, multiplicity, and situated knowing, acknowledging the importance of rigour and precision, but also affect and enchantment. By emphasizing the craft in producing, but also reading novels, and the importance of subjective corresponding relations, the project worked to not only change the positionality of the academic but also what literary geographies can be. Through directly activating an assemblage of author, academic, and audience, each of these roles has become part of a choreographic circuit, with purpose, expertise, and authority transferring and entangling, as the clear categorization of output or process becomes blurred and interchangeable. In identifying the power of this circuit of relations, the second half of this book (Chapters 8–12) seeks to evaluate its potential. Chapter 8 starts with process with an evaluation of how authors engaged with our notion of plot – asking our novelists (and acknowledged experts if the novelists were deceased), how useful or resonant was a sensitivity to the co-ingredience of drama and geography to their novel writing. The following chapters focus on readers. Chapter 9 focuses on how reading groups adopted a plotted reading approach and what differences this made to their engagement with and production of fiction, before Chapter 10 explores how these readers experienced the generating power of the *Literary Atlas* maps more broadly. Chapter 11 documents how readers enjoyed excursions into their borderlands through physical trips to the corresponding geographies of authorial plots, before Chapter 12 engages with readers' experiences of the *Literary Atlas* maps in public exhibitions and workshops. Chapter 13 concludes the book with broader reflection on the *Literary Atlas* in terms of its approach, its evaluation, and its significance for literary geographies in general.

Notes

1 This idea became converted with the skill of Amy Sterly into "The Missing Piece/Y Darn Sydd ar Goll", which ultimately the author of Pigeon Alys Conran felt to "uncannily mimic the geologic quarry around Bethesda" (cited in conversation).

2 We became sensitive to the risks involved in our experiments, "becoming yet another, albeit practice-based, example of the . . . trend towards 'research tourism'" (Hawkins, 2015:249).

3 As the student team goes on to account: "This is not the only commentary that the art piece provides. It also offers commentary and critiques of the novel. We have tried to engage creatively with the themes of exclusion of women within the text by writing women into the narrative, recognising and highlighting the motives and opportunities that exist but go unremarked upon within the novel. This reappraisal of the text seemed necessary considering the outdated treatment of women by the text which keeps them in the margins of the plot and only focuses on women when they have a relationship to important male characters. The text portrays women mainly via accounts of gossiping and descriptions of their bodies . . . . we have re-written the place of women by giving them a place on the murder board in the form of information plaques. We have also attempted to give voice to a critique of the text and the fictional investigation, which overlooked female characters' motives and opportunities to commit the crime, by not creating 'mug-shots' of female characters to make conspicuous their absence from Gallie's narrative" (Savage et al., unpublished).

8 Evaluating the plotted approach

Authorial engagements

As we have seen in the book thus far, the *Literary Atlas* emerged under the influence of a relational sea change across the social sciences, arts, and humanities, a change which mobilized currents of culture, space, cartography, and creativity and offered significant possibilities for doing literary geographies differently. Operating in this confluence of ideas, ambitions, and aims, *Literary Atlas* adopted a plotted approach to engage with, and subsequently map, 12 novels written in English and located in Wales. This plotted approach re-positioned the drama and geography of the story as interdependent in nature, and insisted upon the ontological equality between the author's and reader's worlds that emerged through a plotted reading. From this emphasis, distanced and deep maps were produced, combining authorial locations definitive of the stories, as well as the corresponding geographies identified by the *Literary Atlas* team.

This book would argue that this process and these outcomes are significant. Emphasis on the site of engagement (in the sense of the embodied reader), and how a particular new emphasis on this reading (the plotted approach), can produce new understanding of how we relate to, and indeed configure, the world. This emphasis on the site of engagement also has a second meaning – the production of a new site, www.literaryatlas.wales, to encounter the interpretations conjured through our plotted readings, the compendium of visual representations, maps, images, and words, assembled together to stimulate new perspectives and connections on the literary geographies of the novels in the broadest sense. For the *Literary Atlas* team, the plotted approach was stimulating and intellectually compelling, mobilized through an interdisciplinary perspective that was conjured from beyond the narrow line joining human geography and literary studies, and expanding what an academic is and could be. However, we wanted to not simply present this approach to literary geographies but also evaluate it. Aligned with the commitment to rigour that was integral to our working definition of literary geographies (see Chapter 3), an element of critical evaluation has been an ongoing component of the narrative to date. Notwithstanding this commitment, we were also interested in how our approach may be generative for other "readerships". Thus, evaluation here refers more specifically to engagement and reflection by critical scholars and reading publics, by users of public space (including, for example, reading groups, cafés,

DOI: 10.4324/9781003321248-8

galleries, arts venues, and exhibition spaces), walking tours, and workshop participants. We were committed to asking: what might the plotted reading approach "do"? and what new insights or interest in novels, maps, and localities may be produced as a consequence?

In order to answer these questions, we set about a programme of evaluation "works", comprising a number of stages. Initially, we focused on reflection and engagement with the shortlisted novels, and this occurred in two parts. Firstly, we centred on how authors engaged with our notion of plot – asking our novelists (and acknowledged experts if the novelists were deceased) how useful or resonant was a sensitivity to the co-ingredience of drama and geography to their novel writing (and this chapter critically explores the responses to this question, see below). Secondly, we focused on readers, recruiting established "book groups" to explicitly adopt a plotted reading for two of our shortlisted novels, then discussing with them how such a sensitivity changes their approach to the novel and their understanding of their corresponding geographies (see Chapter 9). Following our engagement with the novels directly, we then sought engagement with the distant, deep, and plotline cartographies produced by the *Literary Atlas* website itself. To this end, we recruited experts on the texts to engage with and review the materials and multi-model maps presented on www.literaryatlas.wales and offer their critical opinion on their accuracy, interpretation, and interactivity. We also surveyed our book groups, following their engagement with the novels, to ask them what new perspectives (useful or otherwise) the website offered to their plotted understanding of the novels and corresponding geographies (and this reflection is presented and discussed in Chapter 10).

In this way, we engaged with evaluation of not simply a plotted reading of the novels, but also the *Literary Atlas* website itself. However, as the role of corresponding geographies had also become central to our focus, we resolved to run a number of public tours based on (some of the) corresponding geographies that had emerged for each novel, in order to evaluate how these places activated the literary geographies of the novel and website in different ways. Oriented around public walking tours, we enacted aspects of the plotlines generated for the *Literary Atlas* website, and the details of their operation and insights are outlined in Chapter 11. The final stage of evaluation aimed to engage with the broader publics' interaction with the *Literary Atlas* in aggregate. To that end, we implemented a series of public exhibitions and workshops in 6 gallery spaces across Wales to exhibit the *Literary Atlas* materials, with a primary focus on the commissioned artistic cartographies (discussed in Chapter 7). Chapter 12 presents the details of this engagement, centring around audience feedback not simply on the curation of these exhibitions but also the accompanying public workshops in which participants could create their own maps (of any sort) in response to the *Literary Atlas* materials. Chapter 8 continues by reflecting on how authors engaged with our notion of plot, exploring the utility of the notion that drama and location can be irrevocably intertwined to contribute to the successful production of the selected novels.

Authorial plots

Given the influence of the current of culture mobilizing an array of social sciences, arts, and humanities disciplines (see Chapter 3), where once a book as artefact may have stood in isolation – as the final comment by its author on the issues it engages with – this cultural trace is now seen as a new beginning, setting in motion an array of eventful engagements which can usefully include not simply the book's audiences but the author's reflections too. In this context, it becomes insightful to ask the author about their intentions for the book and its meanings; whether, for example, it has been received in the manner they had hoped for, as well as inquiring as to their motivations and changing perspectives on the book as time has passed. As Chapter 3 outlines, such activation of the author as a situated agent, intentionally crafting words to set in motion the reader's "architecture of imagination" (Larsen, 2018:168) and corresponding geographies, brings with it an interest in the contexts that influenced the text. For the *Literary Atlas* project, this focused in particular on questions of plot – in other words, asking the author about the importance of the reciprocal significance of "drama" and "geography" in their composition, and thus testing the working "hypothesis" that their story would not be the same story if it were set anywhere else (and by extension, its geography would not be same if the drama was not set here).

Through discussion with authors, it was clear that the relation and indeed interdependence between drama and geography was intrinsic to how their stories emerged. For many, location directly imbued their drama, with the multiplicities involved in the notion of place (including culture, language, socio-demographies, industry, religion, and community), all influential in defining their narrative. Conran, for example, sums up this interdependence at a variety of scales with respect to her novel *Pigeon* (2016):

> The book is Welsh in its very essence, with the brooding atmosphere of the post-industrial slate-quarried environment surrounding the town where it is set ever present, and the web of community (including chapel) and the Welsh language framing almost everything that happens.
>
> (Conran, in Rees, 2016: no page)

Such authorial acknowledgement therefore recognized the importance of place to the development of drama, affirming that an awareness and broader literacy of a particular location enables the creative process, as Fflur Dafydd notes with respect to *Twenty Thousand Saints* (2008):

> I think it must be that the [location], you know, lends itself to the story so that you have the specific places that are fascinating to you as a writer and then you start to devise the scenarios that can happen within them and certainly I think without actually having that experience of being close up to all these locations those stories would never have come . . . you definitely feel as if it's the landscape driving the story rather than vice versa.
>
> (*Literary Atlas* interview with Author)

It was clear that authors gained a literacy of a place's potential for their story in a range of ways. For some, this literacy was based on contemporary experience (Dafydd, for example, stayed on Bardsey/Enlli for a summer which informed her writing of both *Ataniad* (2006) and *Twenty Thousand Saints* (2008)), or drew on reminiscence of the past first-hand experiences (Pryce, for example, based *Aberystwyth, Mon Amour* (2001) on the geographies of his youth growing up in Ceredigion[1]). For other writers, extensive research was explicitly undertaken to inform the generation of their narrative. As Alan Garner discusses, he "found" the plot of *The Owl Service* (1967) through staying in the valley of Dinas Mawddwy, North Wales. Garner was literate of *The Mabinogi* and knew its original locations, however he chose to conjure his own reinterpretation of the story allowing it to grow from this location, a place where he sensed that "fearsome hills hover ominously" over the human settlements below (Hermon, 2006:no page). This sensation he felt resonated strongly with the themes he read in the Blodeuwedd branch of the ancient Welsh myth, as he explains in *The Voice that Thunders*, and then in an interview:

> I set *The Owl Service* in the Mawddwy valley. This was because a friend of my wife had inherited Bryn Hall in Llanymawddwy. It had not been used as anything other than a holiday house since 1898, yet it was still looked on by the valley as the place where the people of power came. My wife's friend wanted the house to be used, and she offered it to us for a long holiday. The concept of *The Owl Service* had been with me for three years, but it had no context, until I saw the Bryn. I had to do the research, but here was the place.
>
> (Garner, 2010:103)

The research Garner then undertook was explained through conversation:

> As I entered the valley, [everything] just fell into place. . . . As soon as I'd got the place and the idea, . . . I [became] a deep mapper! . . . I got the 25 inch map. And there were some marvellous 25 inch maps of that area where there's nothing except little tufts [laughs]. So, I read the maps. I just sank into the maps, and I've always loved maps and I had the ability, even if I hadn't been to the place, to interpret it, but having been there, all I really needed was the map and I could recreate it. I'd also taken enormous amount of photographs and the room where I work, I printed them up on black paper and just lined the room with them: [it was] oppressive [laughs]!
>
> So I did the geography, I did the geology, I did the hydrography, I did the climatology, I did the history, I did the social and political history too so that I could understand the patterns of the land and the fields and so on and so forth and I just immersed myself in it. . . . I knew from experience, I had to let it gestate . . . [it] took five years.
>
> (*Literary Atlas* interview with Author)[2]

Yet Alan Garner did not simply rely on "academic" knowledge to gain an insight into the culture and history of the community. Once at Bryn Hall, he was befriended by Dafydd Rees Clocydd whose family had been in the area for 150 years, and had been keeping the Hall and its grounds since 1898. As Garner explains,

> There was Dafydd Rhys who had been . . . this was 1962 to 3, 1963, and he'd been working at the house since 1898 and looking after it . . . .He gave me everything . . . Within 12 hours of [our arrival] he was talking to me. Now, I cannot explain that. And the result of that [conversation going so well was] over time he just gave me the valley [by sharing its histories and stories from someone who had lived them from the inside].
>
> (*Literary Atlas* interview with Author)

Through his friendship with Dafydd Rees Clocydd, Garner met a range of families in the village and began to gain insights into how the community worked and what the lived experience of the area was like. This intimacy, along with his deep mapping gestation, afforded him a perspective for *The Owl Service*'s plot.

Principle to the plot of *The Owl Service* is a sense of remoteness, both physical and cultural, and this central geographical characteristic is shared with a number of the selected novels in *Literary Atlas* (for example, *Revenant* (Hughes, 2008), *Twenty Thousand Saints* (Dafydd, 2008), and *Aberystwyth Mon Amour* (Pryce, 2001), to name three). This marginality is often directly literal in nature, premised on a physical geography of remote valleys, mountainous terrains, islands, or peripheral towns (at least in terms of the density of transport links to and from it, or the distance from large population centres). As Tristan Hughes notes with respect to *Revenant*:

Q: I was just going to ask about the location, do you think that is integral to your fiction?

A: Yeah, I think it is. I think I'm always interested in remote places or places that are somehow cut off, and, . . . there may not be a sea, there may not be straits, there may not be a definite physical kind of boundary, but they [my plots] are always premised on a place that is remote and cut off in some ways; I think that just speaks to me as a writer. (*Literary Atlas* interview with Author)

In *Twenty Thousand Saints*, remoteness is intrinsic to the geographical location, with Bardsey/Enlli both physically and culturally cut off from the mainland by Bardsey Sound. This physical, fluvial barrier, as Dafydd explains, was influential in plot development:

> What I felt was that the sea itself was giving me the story . . . because there are occasions when the boat doesn't come or when the boat can't cross back over and suddenly what you've got is tension and drama and it totally gives you that. When people are trapped there is the idea of isolation, the idea of people not being able to get across, and wherever I was at a crossroads in the story of *Twenty Thousand Saints* I was thinking how can I heighten the drama again I just knew, right, well okay, the boat won't come and then something

> will happen; so you definitely use the landscape and all its different structures and all its different rhythms in order to drive your story so, you know, without Bardsey giving me these things there would be no story.
>
> (*Literary Atlas* interview with Author)

Due to the influence of remoteness on the plot of our selected novels, it may be tempting to suggest that the stories became wholly premised on an inward, isolated, and perhaps even sedentary sense of place (see Chapter 3). However, as implied in Dafydd's account (earlier), this is not the whole story; a more persuasive interpretation might be that this remoteness and peripherality sensitized the plot to the means through which connection to the wider world was possible and realizable in practice. As Conran puts it:

> *Pigeon* is such a Welsh book but that doesn't mean it's not in a conversation with the rest of the world, you know it fundamentally is in fact.
>
> (*Literary Atlas* interview with Author)

As noted in Anderson (2010, 2015, 2021), places are never purely local (or remote) in nature. Despite the impression that places may be cut off or bounded, it is impossible to insulate a place from increasingly non-local, and often globalized, networks of influence; indeed, it is precisely the composition of these connections that defines the nature of all places. Massey confirms this view, suggesting that

> Instead . . . of thinking of places as areas with boundaries around [them], they can be imagined as articulated moments in networks of social relations and understandings, . . . where a large proportion of those relations, experiences and understandings are constructed on a far larger scale than we happen to define for that moment as the place itself, whether than be a street, or a region or even a continent.
>
> (Massey, 1994:153)

According to Massey therefore, there is a "constitutive interdependence" (2007:21) between places; every place is an ongoing mix of influences that "meet" in a certain location, becoming defined "precisely through the particularity of [its] linkage to that 'outside'" (Massey, 1994:155). Thus plots which appear to be bounded and cut off (e.g. *The Owl Service*) are therefore better understood as composed in unique assemblies with other places, and it is the nature of that circuit of connectivity – the composition of those (b)orders – that offers possibilities for the drama. As Hughes reflects,

> Do you know I like the idea of being able to look at a kind of bounded geography or a bounded place [because really] it's also radically unbounded as well. You realise that the more you look or the more explore an apparently bounded place it begins to open out; there's little bits of history or little bits of a character's life and these begin to take you on journeys or explorations that

> always end up somewhere else. I've always found it fascinating when people think of bounded space it's almost always a fabrication, it's almost always erroneous because what you find is that places are always criss-crossed by other places, they always lead somewhere else, there are always histories that put them in connection with other places, with other times and I think an island [for example] is a great way of seeing that.
>
> (*Literary Atlas* interview with Author)

The (physical) geographies of a place therefore offer dramatic opportunities for an author, whatever location a story emerges from, it is mobilized by that place's (b)orders and the broader circuit the key actors identify with. Each plot, therefore, exists on a sort of frontier for the actor experiencing it, connected and disconnected from other neighbouring places and ways of being. This is most obviously the case in novels such as Amy Dillwyn's *The Rebecca Rioter* (1880). Here, the local protagonists are faced with toll gates taxing their literal movements as they cross the Gower, from remote farms to the markets in Killay. This plot gives a sense of the last days of one way of life as a "civilizing" frontier is expanded into new areas, as Kirsti Bohata acknowledges:

> Yes, there is that there, [it plays into] a tradition – historical as well as literary – of places that are outside law and order, usually on the edge of some wide open space and it is a wild west sort of lawlessness, self-regulating and poaching being the most lawless element of it which Amy Dillwyn . . . puts in her novels quite a lot.
>
> (*Literary Atlas* interview with Authorial Expert)

Physical and cultural geographies, and the nature of their changing (b)orders therefore provide the fertile territory for the dramatic component of plots. An explicit example of this is Raymond Williams' *Border Country* (1960) whose narrative is driven in part by the encroaching (b)orders of twentieth-century industrialization on the physical and social borders of South Wales. As Dai Smith and Daniel Williams discuss:

DS: In a novel of the late 50s set in a place like Pandy it's clearly about the border [between England and Wales] but the question in some ways is the status of Wales within that. Matthew talks about pigheaded Wales. If you're going to put a scene set in an Eisteddfod in a novel, you're dealing with cultural difference [too] in a very self-conscious way. It's always striking to me the way the people occasionally say, "there's nothing very Welsh about *Border Country*". If any Welshman reads it they think actually it's kind of pretty intensely about that . . .

DW: Yes, it is a irreducibly Welsh novel. I don't think it could have been written by anybody else in any other circumstances, its geographical location is always the jumping off point for questions of morality and what is the good life. (*Literary Atlas* interview with Authorial Experts)

The nature of a place's changing (b)orders thus, as Daniel Williams suggests, becomes the driver for *Border Country*'s – but also often many novel's – conversation of what might be a (or the) "good life": how might we value community (e.g. Meredith, 1988; Gallie, 1959), culture (Griffiths, 2001), and the nature of a nation (Dafydd, 2008; Williams, 1960, as examples). In these ways, the coming together of the culturally- and physically geographical become the mobilizing factors in generating the tension and drama of plot. As Conran confirms:

> you can't divorce culture and place. So I think Cher asks Iola what does it matter if Pigeon is no longer speaking to her in Welsh [for example] because Iola speaks to Cher in English, so what does it matter? . . . Iola thinks that it does matter, whole places worth of things [are invoked through its use]. It matters, the weight of it. You weigh how much it matters with place, and a sense of place . . . the language is the medium through which to express the material life, the kind of physical life, and the life of the landscape. So this novel exists because I'm *not* writing in Welsh actually. I think the whole story, the relationship between the characters, what Pigeon comes to kind of mean and to also experience would not have happened if I was writing in Welsh because there would be no tension between the medium of the writing and the place and the characters.
>
> (*Literary Atlas* interview with Author)

From this basis, for our selected writers, how the characters of their fiction interact with their changing physical and cultural (b)orders becomes key. The characters' orientation – or more accurately their *disorientation* (see Anderson, 2015) – in relation to the changing geographical (b)orders becomes the dramatic points of intrigue in the story. As we have seen in Chapter 3, the spatial current across the social sciences, arts, and humanities acknowledges the indelible relations between people and place, as Casey puts it, there is a "constitutive coingredience between people and place" (after Casey, 2001:684). As such, when (b)orders change (either through new societal trends, technological inventions, or other cultural practices), or individuals move away and return to their old haunts, the incumbent need to reorient oneself in new person–places relations is set in motion. In this context, as I have defined elsewhere, disorientation refers to a geographical and cultural process that "temporarily . . . fragments the consolidated rhythms of normal life" (Anderson, 2015:2). Returning to a place that was once considered familiar, or arriving in a new place, can thus disrupt or reconfigure (in other words, disorient) the socio-spatial relations which define a human. How particular characters remain disoriented (for example, in Hughes' Revenant (2008)), fight to retain their person–place relations through resistance (for example, Griffiths' Sheepshagger (2001)), or strive to build a sense of orientation they have never enjoyed (for example, Conran's Pigeon (2016)) become defining dimensions of *Literary Atlas*' selected novels.

For some plots, the role of disorientation to the drama is explored through the explicit practice of homecoming (for example, Matthew/Will's return from London in *Border Country*[3]; Delyth and Vic's return to Cardiff in *Twenty Thousand Saints*; Dolores' return to Butetown in Azzopardi's *The Hiding Place* (2000); Jack to the South Wales valleys in *Shifts* (Meredith, 1988); and Steph and Ricky to the North Wales coast in *Revenant*). Reflecting Eliot's *Little Gidding* (1943), in these plots, key characters return to places where once they had a stable sense of person–place relations, and as they do so they see it "as if for the first time". Through conversation, this process of disorientation (and subsequent reorientation) was also identified as central to the writing process for the authors themselves. As Hughes states,

> the stimulus for me to start writing fiction, or to start publishing fiction, had come from my return here after I'd finished university . . . I think that impulse when I came back and somewhere that I thought I knew very well and I'd kind of grown up in suddenly seemed different and I think that seems to be a common thread really that often a kind of homecoming can kind of stimulate this imaginative encounter with a place that you've always taken for granted. And that returning to it suddenly reveals kind of new dimensions, I guess often new dimensions of yourself; you register the change in your own self, your own set of perceptions, by judging them against a place or a landscape that you're familiar with and then find that it's not as familiar as you thought.
>
> (*Literary Atlas* interview with Author)

Orientation and corresponding geographies

We have seen how coming to terms with, or orienting oneself to, person–place relations is both a key dimension to drive the drama within a plot and also in coming to craft a plot itself. As in the case of plotted reading, where the reader's "architecture of imagination" (Larsen, 2018:168) comes into assembly with the author's crafted words to create a set of corresponding geographies which become irrevocably associated with the plot, the influence of wider corresponding geographies is vital to the author too. For Conran and Pryce, their novels set in Wales are indelibly connected to places they experienced beyond Wales. For Conran, for example,

> What was actually there outside was feeding into this imaginative landscape that I was making which although it is based on Bethesda in some ways, it's also based on other places and it incorporates an amalgam of places and letting it be free in that way was very important to the writing process. [You see] the reason I wanted to write about the place [Bethesda] was I think was because I was finding a home [again] here, having returned from living away. I was finding my way back to Wales.
>
> (*Literary Atlas* interview with Author)

As she elaborates,

> In a strange way I came back through Barcelona – I learnt Catalan whilst I was there and I saw this minority culture at the centre of this cosmopolitan place,

> and people really respected where I was from there, my Welshness, in a way that wasn't the same elsewhere – it just hadn't been acknowledged. There was a tension when I was away until Barcelona – where they saw my being Welsh very differently – and I realised I wanted to come home . . . . I [realised] could not live without the culture and language that now surrounds me.
>
> (Conran, in Rees, 2016:no page)

Just as Conran's experience of Catalonia came to propel her compulsion to write about Bethesda, the distance Pryce experienced from his childhood home enabled him to compose his own version of Aberystwyth:

> I wrote three of them [the *Aberystwyth, Mon Amour* series] in Bangkok, for example, and people always said to me, "how can you write about Aberystwyth living in Bangkok?" And the answer to that is, very easily. In fact, being in the real Aberystwyth I always found was sort of a hindrance in a way. It got in the way of the poetic Aberystwyth, and I think Hemingway said, "to write about a place, you have to have left it first". I'm not sure that's an exact quote, but you can't write about a town until you've left it, and although plenty of people do write about towns they live in, I know exactly what he meant.
>
> (*Literary Atlas* interview with Author)

Thus in the same way that every place is defined by its (b)orders to elsewhere and elsewhen, every plot is defined – at least from the author's point of view – from a variety of person–place relations which have played a role in bringing it into being. As in the case of plotted reading, in many cases, these corresponding geographies originate in the physical and material ontological plane, and in others, they are places originating at least from the author's own plotted reading. For Dafydd, for example, her Enlli/Bardsey was significantly influenced by a version of the place conjured by Brenda Chamberlain in *Tide Race* (1962), as she explains,

> with *Atyniad* I didn't really go back and do a lot of research and look into different aspects of the island. I was banking more on my own emotional memory of the whole period in writing that. When I wrote *Twenty Thousand Saints*, I was doing extensive research into some of the saints that were meant to be buried on the island and the archaeological work that had been done there. I was looking at (really boring) reports about archaeological digs. I was mapping the place looking at the various maps that had been created. . . . I was [also] very taken by Brenda Chamberlain's quote about, you know, "life on this, as on every small island, is governed by the moods of the sea, its tides, its gifts, its depravations". I was reading *Tide Race* over and over so actually I was responding to other texts in a way when I was writing *Twenty Thousand Saints* and thinking, how does this fit in with all of these?
>
> (*Literary Atlas* interview with Author)

For Conran, the literary geographies of Caradog Prichard directly influenced her own composition:

> The way I see Bethesda is fundamentally influenced by Caradog Prichard . . . *Nos Ola Leuad* I absolutely admire and love so much. I think if you've read it and you live here there is just no way that you will not deeply be influenced by it and there are things I think when I look back now there are things I didn't realise were so influenced; Pigeon's mother for one, you know evidently is drawn from Caradog Prichard so there is no question of that. I think the . . . I mean it's difficult sometimes to separate what is drawn from Caradog Prichard and what is just drawn from Bethesda because he was drawing on the same places at a different time.
>
> (*Literary Atlas* interview with Author)

Conclusion

It is clear that adopting a plotted approach offers particular insight into the authorial process of the novels selected for the *Literary Atlas* project. Through conversation, authors and experts were able to discuss the important relationship between the drama and geography of their novels, to such an extent, that they could be understood as being co-definitive in nature. As Dafydd put it, it is as if the place "lends" the author the story, with the location's characteristics (both cultural and physical) driving the story, so much so, that the novel could not exist in the same way anywhere else. As a consequence of this recognition, authors such as Alan Garner go to great lengths to ensure they "have the geography right" (in the words of Trezza Azzopardi (*Literary Atlas* interview with Author)). By harnessing personal experience that has been "banked" (after Dafydd, earlier), or otherwise investing resources into research, authors create their own sense of person–place relations which can create fertile ground for the germination and growth of a plausible drama (which they hope, in turn, will resonate with reader's own personal person–place relations). As Azzopardi puts it:

> For me, it's like a bedrock because . . . partly because when I've . . . and I don't want this to sound too arty or anything, but . . . if I've got my geography right, then that's fine. As long as I know the landscape, I can let people wander around in it. And I know other writers work the other way and they say, as long as I know what this character is going to do, and you know, all the events that happen, then I can set it on the moon. I can't work that way. I have to have my landscape perfect . . . I have to get the geography right.
>
> (*Literary Atlas* interview with Author)

Thus, for the *Literary Atlas* authors, is it as if they recognized and responded to their own person–place relations, and used this awareness to orient themselves to the geographical location of their story. This orientation may be to the physical characteristics of the place or its cultural development, but both components come together to "drive" (after Dafydd, earlier) and define the possibilities of plot.

As we have seen, the cultural components of a particular location may refer to, for example, its language, or its industrial development and decline (and the conflicts emergent from these), but may also refer to the existing literatures that have come to define the area (e.g. *Tide Race* for Bardsey/Enlli). For our selected authors, each of these "real" and "imagined" cultural aspects of a location are influential, and do not appear to be separated or distinguished, in the modern tradition, by them in any meaningful way. As a consequence, the plots as they develop are influenced by a set of corresponding geographies that are, as the plotted approach suggests, ontologically co-equal, and in turn, produce a literary geography which also has the potential to be so for the subsequent reader. Lloyd Jones recognizes this ontological co-equality when he reflects:

> Now, let's not get too metaphysical about this, but we'll talk about a base reality [and holds out one hand] . . . .And here [holds out his other hand] is all forms of virtuality. Now, this virtuality takes in drama, film, theatre, internet, everything that is not of immediate reality. There is an area of crossover though, a fairly large area of liminal area of crossover in between. Now I've always enjoyed being in that area. *Mr Vogel* is very much a book of that liminal area. What I'm fascinated by is the drag, if you like, when we spend more and more time in this virtuality, and how this virtuality is constantly dragging at reality. I'm interested in this liminal area because I spend most of my time in there.
>
> (*Literary Atlas* interview with Author)

Here, Jones recognizes how the modern world categorizes and distinguishes the "real" (on one hand) and the "imagined" (or to use his term, the "virtual" on the other). Jones himself and the plot of his novel *Mr Vogel* emerge from and exist in the "liminal" space in between these two categories. As Jones states, he feels as if he himself inhabits this space much of the time, and how it might be possible for others who spend much of their time reading (or perhaps engaging in other "virtual" pursuits) to do so too. The consequence of this, he feels, is that the modern borders between these two states may "drag" and "merge", producing one "liminal" reality. It could be argued that Jones' articulation strongly resonates with the plotted approach adopted by the *Literary Atlas*, with plotted writing drawing on both "real" and "imagined" geographies to produce plots that are co-defined by both drama and place. Such plots emerge from and exist in this liminal space which is equivalent to the "borderlands" identified by the plotted approach (and also noted by Pullman, see Chapter 1). As Jones goes on to state, those who exist in this liminal borderland are like "truly bilingual people who live in one combined language" (*Literary Atlas* interview with Author); providing a homeplace for those who are truly sensitive to the nature of plot and live in one (read- or written-plotted) reality which (con)fuses all aspects of culture and geography. For these individuals, the world is not "reduced to the ordinary quotidian everyday world, to reality" (Pryce, in interview with *Literary Atlas*), but is rather "more real" or "hyperreal" (ibid., also echoing Eco, 1986). These individuals have created a new type of "map" which, as Niall Griffiths puts it, is used to "steer ourselves through life"

(*Literary Atlas* interview with Author), a new form of orientation which helps maintain a sense of self through various person–place (and what might now be better termed person–plot) relations (see also Chapter 13).

As this chapter has demonstrated, the notion of plot can be seen to have a strong degree of traction with the creative processes undertaken by many of the authors selected for the *Literary Atlas*. An understanding of plot directly motivates and mobilizes the dramatic decisions taken by the authors, and helps contribute to the crafted words communicated to the reader. Yet as we have seen in this book, any novel is only partly produced by the author, and it is the reader that significantly contributes to its completion. This book continues by considering how the practice of plotted reading can apply in this area of novel production, and how it can change the ways in which novels, and their literary geographies, can be understood.

Notes

1 As Pryce states, "Very little" of the novel is based on "contemporary Aberystwyth, 'cause really the Aberystwyth as portrayed in the books is distilled from the Aberystwyth of my childhood, and that's not a question of historical dates or events, it's just a feeling, if you know what I mean? A sort of . . . if you think about your Christmas as a child, they all . . . they might have taken place in different places and you have different backdrops, but there's a feeling which abides, and I'd say that it's a feeling of Aberystwyth when I was that age which informs the sort of texture of the novel" (*Literary Atlas* interview with Author).

2 Pryce discusses a similar process, stating: "It's not like you draw a portrait of a town, you first absorb the town into your sort of soul or wherever and distil it or something, and later you draw on that. It's not a direct, 'oh, there's a town there, I'll describe it'. There's a town, then you reduce it to its essence somewhere inside you, and that's where it stays. Then you draw on that to recreate it" (*Literary Atlas* interview with Author).

3 An example of this can be considered in the following extract from Border Country:

> Matthew walked on into the village, feeling himself already altered, so soon after coming back. But the village had altered too, and he saw this more clearly, walking. Before he had got to Cefn more than thirty heavy lorries had passed him, with high, towering loads. For each, after the first, he left the road and pressed into the hedge. He could feel even there the rush of wind as the weight passed. On this narrow road, a slightly widened lane, which the lorries treated as just an odd mile or two on the long haul to the north, there was no separate place to walk, hardly anywhere even to stand, and the few road improvements only added stretches of speed. The houses seemed now to stand in relation to the road, rather than to each other. It was no longer an enclosed village, but a place on the way to somewhere else, as almost everywhere in Britain was coming to be. Many of the trees under which the farms and groups of houses had sheltered were felled, and the walls seemed more open and more naked. The hedges along the road had been clipped short, and in some places replaced by stretches of wire and concrete fence.
>
> Still, as he walked through, a sense of settlement came back. The fields still climbed unevenly into the mountains, and the earth was red in the patches of ploughland. Far up, under the Kestrel, the wall of Meredith's cottage shone white in the sun, and the pattern was still there, over the broad valley, as the names came back: Penydre, Trefedw, Campstone to the east; Glynnant, Cwmhonddu, The Pandy, The Bridge, Panteg to the west. The soft air, as he breasted the hedge and looked down to the curving alder-lined brooks, was still, on his face, the feel of a known country, and still, high up on Brynllwyd, was the ring of the sheep-wall.
>
> (Williams, 2005 edition: page)

9 Evaluating the plotted approach

Reading groups and reading plots

As outlined in Chapter 8, the first stage of the evaluation of *Literary Atlas* and its plotted approach was focused upon reflection on and engagement with the 12 novels shortlisted. The initial aspect of this evaluation focused on the utility of the notion of plot from the perspective of authors and its ability to cast new light on the writing process. As we have argued throughout this book, any novel is only partly produced by the author, and it is the reader that significantly contributes to its completion. The second aspect of this evaluation was therefore centred on readers. In this chapter, we recount the process of recruiting established "book-" or "reading-groups" to engage with our selected novels, and report on how explicitly adopting a plotted reading approach changed their understanding of the drama and its corresponding geographies. Although the notion of plot, and plotted writing, appeared to resonate strongly with the authorial creation of our selected novels (see Chapter 8), it is widely acknowledged that readers may not necessarily share an author's intentions or dispositions. As Childress and Friedkin note:

> Readers' ability to evaluate and draw meanings from cultural texts that may deviate from the intentions of their creators and producers has been a topic of theoretical discussion in literary theory, cultural and media studies, and sociology.
>
> (2012:49)

Indeed, the capacity of readers to bring their own imagination and interpretation to novels is not only central to why the literary "reading game" is so "special" (after Watson, 2000:110), but also why evaluation of it is so vital; all readers as individuals, and as we shall see, collectives (through the social dynamic of sharing thoughts in reading groups) can assemble their own meanings and corresponding literary geographies for all novels, in new and original ways. This chapter begins by discussing the existing literature on the dynamics of reading groups, introduces how *Literary Atlas* recruited groups for this project, and critically appraises the composition and interests of these groups. It then moves on to discuss how these reading groups responded to the practice of plotted readings; it explores not only how a plotted reading emphasized the geographical dimensions of the drama but also how their "escape" into reading enjoyment was premised on the plausibility of narrative

DOI: 10.4324/9781003321248-9

possibilities emerging from the geographical environment established not only by the author's crafted words but also by the readers' corresponding geographies.

Reading groups

Although reading literature appears to be predominantly an individualized process, the practice of reading "socially" is growing (see, for example, Allington, 2011; Allington and Swann, 2009; Fuller, 2008; Fuller and Procter, 2009; Fuller and Rehberg Sedo, 2013), and the social practice of reading literature is predominantly focused around "reading-" or "book-groups". Whiteley and Peplow define reading groups as "collectives who come together regularly to talk about a literary work, most typically a novel" (2021:119). In both face-to-face environments, and more recently on social media platforms, it is commonly accepted that there has been an "'explosive growth' (Long, 2003:19, see also Hartley, 2001) in reading group membership which continues to be evident today (Peplow et al., 2016:2)" (see Whiteley and Peplow (2021:119).

Much research suggests that the cultural groups who read fiction regularly – or the "reading class", as Griswold et al. (2005) label it – may not be representative of society as a whole. Griswold et al. suggest that perhaps only 15% of adults in countries with full literacy are "heavy readers" (2011:23), and this group tends to be characterized as "urban, educated, relatively affluent, white, [and] women" (Thumala Olave, 2021:100). As Allington puts it,

> book-reading is a relatively exclusive activity, and the reading of the most legitimate books, "literary" novels in particular, is confined on the whole to highly educated people in middle-class occupations.
>
> (2011:319)

Despite the practice of reading literature being often identified as a middle-class pursuit, Tekgül (2019) suggests book clubs can nonetheless be understood as "micro public spheres". Indeed, unlike other recent phenomenon in which new public or "town hall" spaces are characterized as degenerative echo chambers, book clubs are reported to be relatively idealized in their nature. It is suggested that reading groups create spaces in which participants are willing to share their own opinions, be informed by alternatives (see Cameron, 2006:67), and work out standpoints through collective discussion and collaborative interpretation (see Long, 2003; Allington and Swann, 2009; Peplow et al., 2016). As Long puts it, the public sphere of the reading group enables deliberations through which participants:

> clarify their own insights and opinions and also . . . integrate the various perspectives other readers bring to bear on the book. Through this integrative process, individuals – and sometimes the group as a whole – can reach new understandings, whether about life or about the text at hand. The discussion itself, then, can be a creative process, for it elicits a certain kind of value-oriented textual interpretation and encourages (through difference and disputation) a clearer

> articulation of partially formulated perceptions and implicit assumptions, whether about a specific book or about a personal experience.
>
> (Long, 2003:187)

As such, reading groups offered the potential to be good mechanisms through which the potential of plotted readings could be experimented with, and subsequently discussed. In doing so, harnessing reading groups as a method extends and expands the assemblage associated with reading, from not simply a process with the potential for imaginative enchantment with individual's own corresponding geographies, but to an open process where that imagination is shared, performed, and affected by engagement with others.

Given the methodological potential of reading groups in general, new approaches to studying their practice have emerged from a range of disciplines which all connect to some degree to the subject area of literary geographies (see, for example, Whiteley and Peplow, 2021; Childress and Friedkin, 2012[1]). These studies have suggested that reading group debate is often focused on the analysis of writing (as suggested, for example, by the sample questions given for reading groups at the end of Sarah Perry's *The Essex Serpent*[2] (2017)), whilst Laffer notes how "research into reading groups has found that discussions are often dominated by the consideration of characters" (2021:2). With respect to characters and their development, these analyses suggest a potential resonance with the approach of *Literary Atlas*, and in particular plotted readings, as participants appear open to merge or at least suspend the ontological difference between make believe and reality through their reading, as Tekgül outlines, in reading groups:

> readers respond to characters in ways that are not too different from the way they interact with people they share a physical space with . . . During discussions with readers, I often heard them refer to characters in their conversations as if they were mutual friends.
>
> (2019:383)

and as Laffer confirms:

> Long's (1986, 1987, 1992, 2003) seminal work on reading groups . . . argue[s] that readers "often respond directly to fictional characters *as if* they were real people, discussing whether they like or dislike, admire or despise them . . ." (Long, 1986:606).
>
> (Laffer, 2021:2)

The propensity of readers within book groups to elide the difference between "real life and the fictional worlds of a literary text" (after Whiteley and Peplow (2021:120) in relation to characters and their development seems to confirm the position of this book that there is something about "the special kind of reading game" (Watson, 2000:110) associated with novels which encourages readers to merge or at least suspend the modern separation of "real" and "make believe".

Indeed, the context of the reading group itself may exacerbate this tendency, as the elision between "real" and "imagined" can help to create space for agreement between those with different opinions, either by diffusing tension, exerting authority, or finding common ground, as Whiteley and Peplow acknowledge, merging ontological planes of the real and imagined is often used in reading groups to

> claim expertise, to engage in play, and to state (and then backtrack from) controversial opinions.
>
> (2021:136)

Thus, whilst it is important to identify that it is not common for participants to foreground discussion on the connections between geography, dramatic arc, and their constitutive combination in reading groups, this willingness to elide what the modern constitution terms the "imagined" and the "real" suggests a plotted reading may be a useful prompt through which to open deliberation further.

Recruiting

In order to explore the utility of a plotted approach for exploring "socially and culturally situated reading practices distinct from those of the academy" (after Whiteley and Peplow, 2021:120) and thus "provide a window on the everyday interpretation of literary texts" (Peplow et al., 2016:VII), *Literary Atlas* invited pre-established reading groups located in Wales to adopt a plotted reading of our selected novels. Recruiting existing reading groups offered key benefits in terms of methodology; they effectively functioned as focus groups in the social science tradition (see, for example, Conradson, 1997), holding pre-established meetings in organized locations, with established rapport existing between actively engaged members of the "reading class". As such, book groups were likely to offer an interested "public sphere" in which plotted reading could be explored. Eighteen groups were recruited, with all group members giving consent to be involved. Participation was voluntary and unpaid, but all members received free copies of their chosen books.

In terms of book choice, *Literary Atlas* offered suggestions to each group with respect to which novel's most obvious geographies might correspond with the location of the group themselves, but ultimately the final choice of book was the group's decision. In terms of their approach to their chosen novel, we would ask each group to do three things. Firstly, in order to set the context for their plotted reading and subsequent reading group discussions, we asked each participant to engage in a survey about their existing reading practices. Informed broadly by the "cultural sociology of reading" approach (Thumala Olive, 2021:101), this survey was motivated to explore:

> why reading matters to people, why readers care for their reading practices in general and for particular texts, what their attachment looks like and what sustains it.
>
> (ibid.)

Secondly, we invited all book groups' members to undertake a plotted reading of their chosen novels, and, thirdly, to meet as a group in their normal way in order to discuss their responses to this reading. Members of the *Literary Atlas* team would attend this discussion, ask questions of the group if necessary, and audio record the conversation for subsequent transcription and evaluation. It was agreed that all contributions to the discussion (in terms of specific individuals and the book group as a whole) would be anonymized. Through this process, each of the *Literary Atlas* novels was read by our book groups, with every book having between 20 and 37 readers. The novels were distributed as follows:

1. ***Aberystwyth Mon Amour:*** 32 readers across 5 groups: Porthcawl Library, Gwaen Cae Gurwen, Goginan Group, Bargoed Writers Group, and Aberystwyth Waterstones
2. ***Border Country:*** 31 readers across 3 groups: Monmouth Writers' Group, Bargoed Writers Group, and Cymmer Afan Library
3. ***Mr Vogel:*** 23 readers across 3 groups: Cardiff Museum Writers, Bangor students, and Pembrokeshire Book Club
4. ***The Owl Service:*** 33 readers across 4 groups: Llangollen Library, Penarth Library, Aberystwyth Students, and Waterstones Aberystwyth
5. ***Pigeon:*** 32 readers across 3 groups: UWSTD Students, Paperwings Bangor, and Llangollen Library
6. ***Revenant:*** 30 readers across 4 groups: Penarth Reading Group, Paperwings Bangor, Bangor Students, and Copper Writers
7. ***Sheepshagger:*** 24 readers across 4 groups: UWTSD Students, Aberystwyth Students, Cymmer Afan Library, and Goginan Group
8. ***Shifts:*** 27 readers across 3 groups: Monmouth Writers' Group, Bargoed Writers' Group, and Bargoed Library
9. ***Strike for a Kingdom:*** 36 readers across 3 groups: Swansea Central, Pontardawe Library, and Gwaen Cae Gurwen
10. ***The Hiding Place:*** 22 readers across 3 groups: Cardiff Museum Writers, Penarth Library, and Cardiff Writers' Circle
11. ***The Rebecca Rioter:*** 37 readers across 3 groups: Swansea Central, Pontardawe, and Porthcawl Library
12. ***Twenty Thousand Saints:*** 24 readers across 3 groups: Pembrokeshire Book Club, Cardiff Writers' Circle, and Penarth Reading Group

Survey

To set the context for our plotted reading discussions, members of each book group were invited to fill in a survey about their motivations, habits, and approaches to reading. Fifty-five book club members opted to complete the survey, and the responses are outlined below.

The socio-demographics of this "reading class" were as follows: 45% were aged between 18 and 35; 11% between 36 and 45; 12% between 46 and 55; 12% between 56 and 65; and 20% over 65. The large percentage of readers within the 18–35 age group reflects the recruitment of three student reading groups for the project. In terms of gender, 43% of total respondents self-identified as female, 16% as male, and 40% chose not to declare any gender identification. In terms of occupation, 23% declared themselves to be retired (slightly higher than the number of respondents of pensionable age) and 56% were involved in "education" broadly construed (14 were students, 1 was a teacher, 2 were lecturers, and 2 were librarians). The remaining cohorts who stated their occupation were in white-collar jobs (including, for example, general practitioners and management consultants). This demographic suggests a broad alignment between the *Literary Atlas* reading groups and the conclusion that "the field of book reading is still overwhelmingly populated by the educated middle classes" (Bennett et al., 2009:106, also cited in Allington, 2011:319). In terms of nationality, of those who expressed an allegiance, 13% identified as Welsh and 24% as British. 94% of respondents read fiction predominantly (as opposed to non-fiction), with 88% reading novels "a few times a week", and 54% reading novels "every day".

We then asked this cohort about their favourite writers. Responding to this open question, a diverse range of 162 different authors were suggested by name, with the most popular being: J. K. Rowling and Jane Austen (7 suggestions); Margaret Atwood (5); Jojo Moyes (4); John Grisham (3); Neil Gaiman (3); and William Shakespeare (3). When asked about their favourite Welsh writers (or writers that write about Wales), 305 different authors were suggested, with the most popular including Dylan Thomas (41 suggestions); Kate Roberts (19); Amy Dillwyn (14); Alys Conran (13); Gwyn Thomas (9); Gwyneth Lewis (9); Alan Garner (8); Raymond Williams (8); Owen Sheers (6); RS Thomas (6); and Trezza Azzopardi (6) (see also Chapter 10).

As noted earlier, research on reading groups suggests that participants tend to focus their discussions on characters and language. In response to an open question regarding reading motivations, these factors remained key for the *Literary Atlas* book groups, as the following responses confirm:

> I enjoy getting to know the characters and the world through their eyes.
>
> (Survey Response)

> I love entering new worlds in my mind and living different lives with the characters.
>
> (Survey Response)

> I like the way a book is written, the themes, the story, discovering new things, as well as getting to know the views and writing styles of the authors.
>
> (Survey Response)

In terms of authorial language and craft specifically, the respondents stated:

> I like finding out how the authors describe events and places. I like to dwell on particularly striking phrases and descriptions. I am interested in how language is used.
>
> (Survey Response)

> I mainly like the use of language – arresting phrases. It's like borrowing other people's minds. It changes your perceptions.
>
> (Survey Response)

Whilst these responses solidify the conclusion that reading groups are motivated by engaging with, and subsequently discussing, character and written style, if we explicitly adopt a plotted lens to these survey responses, we can also bring into focus new motivations for social reading. As the first set of responses suggests, there is an element of "discovery" and a desire to "enter . . . new worlds" which drives a continual return to reading. Following Chapter 3, if we embrace the human geographic understanding that posits a constitutive relation between people and place – in other words, that we are not defined simply by aspects of liberal individual selfhood in isolation but rather our humanity is defined as part of an assemblage of social–spatial relations – then reading can be understood as part of a need to experience change in the person–place relations that define a person. The pull of "living different lives" and entering new worlds, can be interpreted as adding new dimensions to readers' existence, involving them in new assemblages of places, cultures, and characters. Explicit mobility from one place to another most obviously disrupts existing person–place relations, and the act of enjoying, and perhaps even needing, such disruption of person–place relations is termed by Anderson and Erskine (2014) as "tropophilia" – defined directly as the "love of change". Such tropophilia is often most commonly associated with physical travel (see also Holton and Riley, 2014), however, disruption in person–place relations can also occur when places themselves (inevitably) change through development or decay. In relation to our respondents, however, desired disorientation was not mobilized through physical travel, but through engaging with a book, as one respondent puts it, "reading is the possibility to travel to other places *without* moving" (Survey Response, emphasis added). Through the practice of reading, it is possible for an individual to transport themselves into a new "read-reality" through the crafted words of the author, and the corresponding geographies conjured to complete them.

This notion of reading in order to actively augment person–place relations and pursue identity challenge and transformation (see also Rojek (1993) is explicitly noted in the following book group responses:

> I like escapism, something that takes me away from modern life.
>
> (Survey Response)

> I read to get away from everyday life.
>
> (Survey Response)

> [I like] escaping, learning, encountering new experiences, imagining characters, visiting new areas.
>
> (Survey Response)

> I read as an escape from real life.
>
> (Survey Response)

The adoption of the active verb "escape" to describe the experience of reading suggests not simply a "push" factor acting on the reader which compels them to do something to get away from a set of habituated, regularized, and perhaps often mundane person–place relations but also a pull factor in escaping to a set of different others – with their identity defined by new customs, rituals, and people. As we have noted, escape is a desired experience and fulfils a need for "disorientation" in person–place relations (see Chapter 8); if and when this need is fulfilled, it can be seen to function in the same way as the practice of the carnivalesque (after Bakhtin, 1984). As Nygaard identifies, the carnivalesque enables "an unhindered interaction between people who otherwise do not interact" (2018:372), in this case, between the reader and types of character they may rarely meet in their everyday lives. The carnivalesque enables an individual to "explore things [one] wouldn't do normally" (2018), for example, different places, histories, and existing within different structured genres, and through the act of reading these co-compose with their normal socio-spatial knowledge, experiences, and imaginations. In sum, when engaging with literature, the reader enters into a read-reality which offers a (temporary) release from the constraints of their everyday (after Bourdieu, 1977, 1991). Reading offers the capacity to intentionally disorientate themselves by re-composing a plotted-self in a "borderland" composed of new characters, relations, places, cultures, languages, etc., which are nevertheless tied to their own. As the following respondents put it, it is an active, stimulating, and liberating experience:

> Books take you to another place where you learn new things both about the content of the book, but also much about yourself.
>
> (Survey Response)

> You forget the world around you and explore a new and different world.
>
> (Survey Response)

> Reading involves me diving into another world; it's relaxing; it helps me to enlarge my imagination.
>
> (Survey Response)

> I dive into the story and love getting caught up in it. I love engaging with ideas and experiences, the way it makes me think and imagine things.
>
> (Survey Response)

In this spirit, the carnivalesque dimension of plotted reading can further inform our understanding of the practice by drawing attention to the ways in which the escape offered is not simply from the quotidian everyday but also its associated modern constitution. As we have seen earlier, many participants in reading groups talk about how they elide the difference between imagined characters and real people, and this merging of ontological boundaries can be seen as another form of escape, in this sense from the modern constitution and what Gill et al. (2018:183–186) term the "the unhelpful 'boxing-up' that a compartmentalized world view can promote".

In this sense, reading fiction enables readers to position themselves within an alternative ontology which "carves up the imagined geography of discrete categories" (Gill et al. , 2018), and re-imposes a new "read-reality" where modern (b)orders no longer apply (as respondents noted earlier, reading is "an escape from real life"). More specifically, escape here can be understood to be not simply into a world of literature but also away from a plane where modern "reality" is reified, and into a new world where this view is subverted, and "make believe" and cognition exist as ontological equals. Resonating again with the notion of the carnivalesque, reading in this sense can be seen to blur the ontological domains of "real" and "imagined" so that a new "layer of reality" (after Stallybrass and White, 1986:6 in Chaabane, 2020:no page) is enabled which can "challenge the social, political, symbolic and moral status quo" (after Chaabane, 2020:no page).

It is possible therefore to suggest that members of reading groups are motivated to participate in reading fiction in order to escape into a different "read-reality", where – prompted by the crafted words of an author – they can build characters and plots in their own "architecture of imagination" (after Larsen (2018:168); as one respondent puts it, they read to: "extend the imagination's boundary and explore the human condition led by the writer" (Survey Response). These characters and plots are populated from the suggestions of the authors, but also inevitably combine with the readers' own experiences, insights, and corresponding geographies (as the following respondents put it, this "stimulating" activity "helps me to enlarge my imagination . . . [and] create images according to the story in my mind, experiencing the character's emotions" (Survey Response)). As a consequence, reading offers an escape to new amodern worlds of experience, of affective disorientation including both "magic and horror" (Survey Response), that is, "relaxing, enriching, entertaining and makes a contribution to my wellbeing" (Survey Response). Following this individual process of "escaping, dreaming, learning, unlearning, loving, knowing, and unknowing" (Survey Response), readers can then become involved in the collective process of "sharing, blethering afterwards, and coming back down" (Survey Response).

Discussing plotted reading – "location-wise": "I think this one is far more, location-wise, you know?"

Thus, from survey responses, it was clear that although reading groups appeared to be motivated by an interest in characters and language, it was also possible to interpret their continued engagement with reading in terms of their wider person–place relations, their tropophilia, and their desire for carnivalesque-style disorientation. Given the resonance of such framings with the plotted reading approach (as well as the plotted writing approach, see Chapter 8), the reading groups were keen to discuss the utility of plotted reading to their understanding of literature and its wider geographies.

Emphasis on plotted reading led the *Literary Atlas* book groups to initially discuss the ability of writers to conjure a convincing and compelling location through their choice of crafted words. As the respondent earlier puts it, they effectively judged whether an author was "location-wise". Although it might be fair to interpret this response in terms of the reading group participant searching (and perhaps momentarily failing) to find the apposite word to describe the detail and resonance with which one author described the physical and cultural geography of the drama compared to another (i.e. "I think *this* one is far *more* location-wise, you know?") – it is nevertheless instructive at this point as it draws attention to the authority commanded by the writer if they can describe a location in an evocative and convincing way (i.e. this author was more location-"wise"). Wisdom in relation to landscape description was suggested to be the way in which, just through a few words, an author could conjure a sense of place for the reader. Raymond Williams was seen as a very strong example of this skill, as the following respondent sums up:

> I did enjoy [*Border Country*] mostly because of the way in which [Williams] portrayed the countryside which he does really well and I mean here's just one [example], but you could take any one sentence:
>
> *"Fields climbed unevenly into the mountains and far up on the black ridges stood the slated houses and the grey barns".*
>
> In just one sentence! I mean I go to rural Wales a lot and that describes it perfectly. He's brilliant at describing the Welsh countryside. At the start of Chapter 10, which I won't read, but it's a wonderful description so to me the quality of this book is based more on this geography than anything.[3]
>
> (Reading Group Participant)

The full quotation alluded to by the reading group participant is as follows:

> Once they were up on the road, Harry and Ellen could look out over the valley and the village in which they had come to live. To the east stood the Holy Mountain, the blue peak with the sudden rockfall on its western scarp. From the mountain to the north ran a ridge of high ground, and along it the grey Marcher castles. To the west, enclosing the valley, ran the Black Mountains: mile after mile of bracken and whin and heather, of black marsh and green springy turf, of rowan and stunted thorn and myrtle and bog cotton, roamed by the mountain sheep and the wild ponies. Between the black ridges of Darren and Brynllwyd cut the narrow valley of Trawsfynydd, where the ruined abbey lay below the outcrop of rock marked by the great isolated boulder of the Kestrel. Fields climbed unevenly into the mountains, and far up on the black ridges stood isolated white farmhouses and grey barns.
>
> (Williams, 2006 edition:36)

This reading group participant effectively acknowledges Williams' ability to use a few well-chosen crafted words to evoke a sense of place which resonates strongly with the corresponding geographies of the reader. The reader is urged to personify fields and, perhaps from a viewpoint across the valley, watch them edge up mountains, clinging to and collapsing over the craggy terrain. Then, standing solid in such surroundings, the farm-building topped with the mined resources from these hills, both imposed but somehow emergent from the land itself.

As the respondent states, Williams' words echo and augment their own experiences of the place – they claim some first-hand engagement with contemporary "border country", and although this experience is 60 years after the publication of Williams' words, their "telling detail" nevertheless have the capacity to "touch [contemporary] life" (after Bradbury, earlier); they create a "borderland" for the reader that is not simply the "*Border Country*" of the novel, but strongly corresponds with the physical geographies of border country today. A similar process of evocative correspondence occurred for many readers in relation to Dafydd's Bardsey Island/Enlli in *Twenty Thousand Saints* (2008); in this case, a sense of location was conjured for respondents who had no first-hand experience of the place:

> although this was a work of fiction, there were places, references to places on Bardsey, and the feel of the place that were conveyed to me through them meant that although I'd never been there, there were things there I could recognise. Like when there's a view . . . of the . . . sort of the guest resident writer that's heading off over to the island and that she has to head north, she's missed the boat and she goes up to this viewpoint and from that viewpoint you can see the highpoint of Bardsey Island and the description of it lying like a "cat in the water". That kind of thing. That brought memories back to me of what a friend who was a warden on Bardsey years ago had said to me when he had to travel across the island, so I think that's quite incredible!
>
> (Reading Group Participant)[4]

The relaxed feline shape of Bardsey Island is depicted on the front cover of Atyniad (Dafydd, 2008, see Ylofa, n.d., see also Figure 9.1) and is referred to the text of *Twenty Thousand Saints* in the excerpt below:

> They both sit up, and look across. There it is. Bardsey Island, lazy in the water. It looks like an outstretched cat, its long green paws leading up to its gorse-carpeted, rising hind.
>
> "It's so small," she gasps, "and so beautiful."
>
> She looks across to what will be her home. Though she cannot see the cottage, the tip of the lighthouse is visible, winking at her. She imagines walking out at night, looking up at it and feeling its greatness, its towering mystery.
>
> "You'll be bored out of your brain," says Mark. His face becomes a shadow once more.
>
> (Dafydd, 2008:20)

In these cases, the author has been able to call forth readers' imagination to project a sense of place of the location in which the novel emerges – this is sometimes achieved through harnessing the readers' first-hand experience of a place, but in most cases does so through mobilizing a set of other corresponding geographies – from other media, from expectation, from assumption, and from pure imagination – to

Figure 9.1 Fflur Dafydd and Kirsti Bohata (Literary Atlas) talking on the Llyn Peninsula with Bardsey Island in the background.

create not simply the literary geographies of the novel, but by extension how the world itself is understood.

A further dimension is introduced into this practice when there is a close alignment between a location noted in a novel, and a reader's knowledge of a place they correspond with it; for example, when the crafted choices of the author call forth specific streets which are known, and perhaps even inhabited, by a reader. This was the case in Niall Griffith's *Sheepshagger* (2001). For some reading groups, Griffith's Aberystwyth resonated strongly with their first-hand experiences of its streets:

I: What did you think about Griffiths' representations of Aberystwyth, the town of Aberystwyth?
M1: Spot on, bang on.
M2: In the town? There were all the pubs!
F1: Several identifiable pubs.
M2: *The Castle*, *The Beach*, *The Fountain*. I've never been in that pub but I'm going to go in it now! [laughs]
M1: I did think some of the descriptions were bang on too, particularly Cyril's shop, and they described Cyril, his shop and everything down to a "T", an absolute "T".
I: And this is an actual shop?
M1: Yeah, some of us remember this guy. He's dead now.
F3: Yeah, and his shop is boarded up now.
M3: It's where Roger buys a dildo!
M1: Cyril tried to sell me dope the first time I went in his shop [laughs].

(Reading Group Participants)[5]

As we have seen earlier, if there is a strong resonance between how a novel's location is described and how it is known by a reader, the "authority" of the writer's words is compounded. Being seen as "location-wise" by an audience therefore emerges from the perceived alignment between the literature's "telling details" and how that location is known by its readers. However, many individuals in the *Literary Atlas* focus groups acknowledge that this alignment was not always possible. As we have seen in Chapter 3, the spatial current has pluralized our consideration of location, requiring us to consider places as emergent, dynamic, and always changing. As a consequence, any reader approaching a novel is not likely to be "looking for [the] one" and only essential version of a place (after Brown, 2018), but is rather aware that any place can be understood in an "infinite" number of ways (after Solnit, 2010). This acknowledgement was implicit within the approach of many reading groups, acknowledging that any location could be known in a variety of different ways by both writer and reader, that different characters would experience a place quite differently from each other, and thus any singular sense of place or specifically accurate locational description would likely be elusive. As the following respondent points out,

> Reading highlights the subjectivity of [knowing] a place . . . and it's kind of exciting really to think about the way that you think you know somewhere and then you might read a novel about it and it's completely different and you suddenly realise that there's another perspective and another perspective and there's all these different ways of looking at one thing. So I think that a novel can provide a new understanding of somewhere and I think you can enjoy their experience of it as well as have your own, you know?
>
> (Reading Group Participant)

This acknowledgement of the inevitability of different authorial (and reader) interpretations of a place thus reinforces the view noted earlier that encountering alternative versions of person–place relations is one motivation for continuing reading (group) practice. Although it was often life-enhancing and energizing for many readers to have their own experiences reflected back to them through an author's crafted words, it was acknowledged that there was also pleasure to be gained from established person–place relations being disturbed and disoriented, as the following conversation indicates:

R3: I don't really care if the version of the place that I'm reading is realistic [when compared to my own experience of it]. It's like, if I feel like I'm connected with this place, even if it's a made up version of this place, then I'm going to go with it.
R4: Because who's to say what's real, right?
R3: Right!
R4: You know, maybe dragons did once exist here in Wales – because I read it in a book, that means that it's real for me!
I: There are definitely dragons. I think there's no mistaking that.
R3: We're pretty sure, it's on the flag!

(Reading Group Participants)

These comments emphasize how plotted reading draws attention to the diverse nature of places that can be conjured through the crafted word, and as long as these words offer "telling details" to the reader, regardless of their technical accuracy, they have the potential to mobilize the reader's corresponding geographies to create a read-reality. This read-reality has the effect of "dragging" (after Jones, see Chapter 8) ontological planes together so it is not useful or desirable to the reader to distinguish between the two.

However, given the multiplicity of place and, therefore, the infinite number of possible dramas that could emerge from one location, the "trick" remains for any author to convince the reader that their chosen dramatic arc could be constitutive of their chosen place. In other words, can they persuade their readers that their plot is convincing? For some participants in the *Literary Atlas* reading groups, this trick was not always pulled off.

M1: . . . some of the [passages in Niall Griffiths' *Sheepshagger*] were incredible of the surrounding scenery but the locations don't actually sort of join up. I mean he [Griffiths] talked about in one part of the book walking down from the hills above Llangurig to the square in Llanbadarn. Well, they're miles and miles away from each other; It'd take you a day or more to walk from one to the other, so that didn't really tie up.

M2: But I don't think that actually mattered, 'cause the description of each place was so . . .

F1: It does matter if you're local [laughs].

M2: I don't think it did.

F3: Yeah, that didn't bother me that much, I have to say . . . .that didn't trouble me from the point view of a local. I kind of went . . . 'cause I don't mind being pushed around a little bit in where I am, 'cause I thought it's a book, or it's . . . a story and that's what you're doing.

(Reading Group Participants)

As we can see earlier, it is possible for authors' words to lose their magic for some readers. The conversation earlier outlines how, for some readers, geographical accuracy remains vital to keep the "magic" of a read-reality alive – not being able to traverse the read-reality via conventional means was thus one way for the reader to doubt the credibility of the author's words. However, for others, discrepancies in details had little effect on the conjuring process. In this sense, what is commonly understood as "poetic licence" was seen to minimize any "jarring" effect for some readers, as the following book group participants put it:

I: Do you think it does matter when, if a novelist has chosen to set a book in a specific place, do you think it's a problem if they then re-shape its geography?

F3: I think artistic licence is okay when the author hasn't mentioned a specific place. Otherwise I don't think you should because it's very obvious where somewhere is.

F3: Yes, because when you know the place, because, it jars.
M1: You start to mistrust them.
F3: Yeah, you start to mistrust other things that are said, then.
M1: Yes. And you think does she want you to believe it's here, or somewhere else, or a nowhere?
F1: And then I find it difficult to get into.

(Reading Group Participants)

Poetic (or artistic) licence in this case does not simply refer to the capacity of an author to make up their own plot, but rather it refers more directly to the authority granted by the reader to allow an author scope to rework a geography to suit the story. This "licence" may be granted by the reader based on how effective the author's crafted words are in mobilizing the reader's corresponding geographies, but also on a reader's tolerance for having their person–place relations disrupted in a particular way.

Culture and corresponding geographies

As we have seen in Chapter 3, the cultural and spatial currents mobilizing the social sciences, arts, and humanities insist not simply that all human life is geographical to some extent, but that this geography is also social or cultural in its nature. Thus, when we consider how effective an author's crafted words are in mobilizing a read-reality for the reader, it is also important to consider how a place's culture aligns with readers' expectations. We can ask, for example, to what extent do an author's crafted words convince the reader that these characters and this culture could emerge from their chosen physical location? Two authors which the *Literary Atlas* reading groups felt excelled in co-composing culture with geographical location were Amy Dillwyn (in *The Rebecca Rioter* (1880)) and Menna Gallie (in *Strike for a Kingdom* (1959)):

M1: So I read the *Rioter* first, the Amy Dillwyn book, and that had huge resonance for me . . . I think what appealed to me about the book was that it was, that there was a sense of claustrophobia about living in small villages. I certainly got a sense of that growing up in one you know . . ., there's a lot of gossip going on and there's a lot of sort of things you know being said and you certainly got a sense of that in this book . . . I find the [book], looking at the situation from my memories, simply amazing. I thought it was really, really, really good. I really, really enjoyed that.

(Reading Group Participant)

F4: [*Strike for a Kingdom*] was beautifully written, it was absolutely lovely, and the character descriptions were so sort of typical of . . . I'm thinking of my grandmother, who'd be sort of 120 now if she was still alive, I lived with her when I was young, and it really captured life in the Welsh valleys, I think that was really good. And the names of the characters, very typical.

F2: Yes, I liked that. I mean you were saying about your grandmother, I remember my great grandfather, and do you know when you were reading it and they were having the discussions with each other, I kind of remembered back to when I was a little child listening to him speaking, and that's how he used to talk. And it does take you back and you do . . . I liked that.

F4: It was very true to life, and it depicted also the closeness that existed in the villages, you know? Everybody else keep out, we'll look after our own and we'll sort ourselves out.[6]

F2: Yes, yes.

F4: And the vocabulary and the turn of phrase, I loved it, loved it.

F2: I did, I really enjoyed it, and like I said, you know, when there were the dialogues between them and they were talking, I could hear my great grandfather speaking to the older people when we were children, we used to go visit him, and it takes you back, it does make you think about . . .

F4: I think she got it to a 'T' didn't she? yes.

F2: When I was reading it, I was in the valley where they were . . . when I was reading it, I could imagine myself being in the street and you know, the neighbours looking out for each other and . . . like with Gwen when she died and how everybody would close the blinds. You don't see that now, you hear someone's passed away and it's, 'oh, I'm sorry' and what have you, but back then, it was nice that everybody would close the blinds out of respect.

(Reading Group Participants)

F5: The village [in *Strike for a Kingdom*] was brought alive for me really. With the claustrophobia of living in such a close community and also the immense support service because everyone is willing to help and you know so that I think was the most identifying factor for me.

F8: One of the words I'd use is intimate. And I think that reflects sort of village life perhaps at that time and again with

F7: Close knit

F8: As a child I remember we ran free in the village. If anyone got hurt there would be a neighbour who'd come and pick you up or whatever . . . and take care of you but also.

F2: [interrupts] suffocating really.

F1: You knew it was a Welsh valley and . . . the language and . . . the stories were real. You knew they were real, exactly as you said, you knew she what she was talking about. It was brilliant. It was a brilliant book.

(Reading Group Participants)

Thus, by adopting a reading which is sensitive to the notion of plot, reading group participants become explicitly aware of the importance of telling details not only regarding the physical geography of a location but also the cultural geographies that each author has chosen to emerge from these. If it was possible for authors to use their crafted words to create a plausible and compelling culture for their plot, readers could sense this authority and it served to bind the plot with the readers' own corresponding geographies creating a sense of affective connection and hiraeth (or wistful affection) for past places, and present plots – for them, these "stories were real".

Discussing dramatic plot

A further key dimension of the notion of plot is the dramatic arc that aligns with its geographical basis, and the importance of this appearing to emerge co-constitutively from it (see also Chapter 8). It is clear that many members of the *Literary Atlas* reading groups felt that some plots were strongly aligned with these geographical locations, including, for example, *Sheepshagger* (Griffiths, 2001), as the following respondents sum up:

R3: Griffiths did a very interesting job of capturing the beauty of the place but then also making sure you're not blind to the sort of more brutal and real aspects of living kind of out in the middle of nowhere where nature is, which is of course another thing in itself. I thought it was good because it does capture that sense of tourists and students, you know, we just see that, 'oh pretty ocean, pretty hills, pretty . . .' you know, but it's not *just* a pretty place, it's a place where life happens and I think that it did a really good job of capturing both things and making you think about the other side of things that you don't necessarily want to. But then you think, well, I should be thinking about them more deeply.

(Reading Group Participant)[7]

Due to Griffith's ability to use his crafted words to set up a plausible plot (in terms of geography and culture) of the Aberystwyth hinterland, it was possible – as one reading group participant put it – for more "gruesome" aspects of the plot to emerge "naturally" (Reading Group Participant). However, not all of the selected novels were convincing for the *Literary Atlas* reading groups.[8] One particular example was Malcolm Pryce's plot for *Aberystwyth Mon Amour*. As noted in Anderson (2014), Malcolm Pryce has always felt that the cultural geographies of Aberystwyth have lent the town an absurdist dimension:

> There was a bus from the village to Aber every morning, but not one back. That is so funny. I feel . . . you can't capture that sort of quirkiness directly; you have to apprehend it obliquely, poetically. (Malcolm Pryce, Personal Communication, cited in Anderson, 2014:197)

From this absurdist geography, Pryce chose to create a detective (comic-) noir, juxtaposing his interpretation of the physical and cultural characteristics of Aberystwyth, and heightening these by adding druids, gumshoe detectives, old-lady office cleaners, and candy cane making (evil) geniuses. The alignment between the two key aspects of plot – geography and drama – was a point of key disagreement between some members of the *Literary Atlas* reading groups. For some, Pryce was cleverly using different literary styles and genres to disrupt one version of Aberystwyth and suggest another. For other discussants, this juxtaposition was so disruptive that they simply did not accept this alternative version could "truthfully" emerge (after Bradbury, 2008:108) from the geographies of their corresponding Aberystwyth. As the following excerpts summarize:

M2: I thought it was a cartoon version of Aberystwyth . . .
F4: This book is like essence of Aberystwyth, it's a pastiche.
F1: It's a parallel universe.

~

F1: I liked it. It made me laugh out loud in parts. It really did. . . . But I mean, it was a satire, wasn't it? Please tell me it was a satire?!
F5: Yeah well I thought it was a bit quirky and I love something a bit quirky.
F4: I really enjoyed it but I mean, poor Aberystwyth, it's a terrible advertisement really. . . . It seems to be the crime centre of Wales!

~

F1: Oh, I thought it was absolutely ghastly.
F6: Yes, my feeling is, I mean Dylan Thomas wrote about Laugharne, but he [gave] the village another name, so he could do what he liked. If you're going to call it Aberystwyth, I think you . . . should have a bit of respect for the location, really. But that's my personal view.
F4: I know it's all a spoof and everything, but I found it intensely irritating. [laughs]. I thought they were making the Welsh look like a load of bloody idiots!
F1: I don't see any defence for it really.
F3: I don't recommended it as good reading.

(Reading Group Participants)

As we have seen in relation to the academic literature on reading groups, individuals who participate in them often create space for others with alternative opinions to their own, and do so by downplaying the importance of their discussion by broadly acknowledging "it's only make believe". We can see something similar at work here, with reading group participants resorting to this modern categorization of the world to enable different opinions to coexist in one space, and the respondents simply acknowledging their differences and accepting that they are *only* disagreeing about "make believe". However, it can also be suggested that in these different plotted readings there is something else occurring. Whilst participants may still acknowledge it is "just a book", they have also sensed how books, and the (plotted) reading of them, are powerful. From their own experience, they are aware that a plotted reading can sensitize and strengthen their connections to a plot, binding it to their own sets of corresponding geographies (as we have seen with respect to many selected novels above). Through undertaking a plotted reading, book club participants are aware, therefore, of the capacity of plotted reading to "merge" (to use the words of Lloyd Jones, see Chapter 8) the boundaries between what is real and what is simply make believe, with the power of plotted reading "dragging" at these differences so that perhaps people no longer maintain them. As one participant puts it, through a plotted reading, "you're no longer sure what's fiction anymore" and, as one respondent put it in relation to *Strike for a Kingdom*, the "stories are [now] real." When this "read-reality" emerges, it is as if some participants become aware that this may be unfavourable when compared to dominant versions of more conventional reality – as one participant puts it with respect to their new read-reality of Beaumaris after their plotted reading of the Tristan Hughes' *The Revenant* (2008)*:*

> *F2:* I think reading this book has changed my perception of Beaumaris, because I think that there must be a cold side I had absolutely no idea about. I've always loved it in the past. I've always thought it was a lovely place. . . . I will never look at Beaumaris in the same way again.
>
> (Reading Group Participant)

Thus, through plotted reading, book club participants became aware that "read-realities" could no longer be dismissed as simply "make believe"; plotted reading had activated a new power to create literary borderlands in which to escape, and live.

Conclusion

As we have seen in this chapter, plotted reading draws attention to the important role of geography in drama, and that engaging in this practice is understood by many as a form of escape. Echoing Rojek (1993), reading plots offered book club participants the opportunity to escape their everyday person–place relations, and enter into new sets of cultural and geographical relations which can broaden, deepen, or otherwise disorient their norms. Plotted readers therefore appear to enjoy this experience – perhaps defined by the term tropophilia, or love of change in person–place relations (after Anderson and Erskine, 2014). As we have seen, it is important for any author to convince the plotted reader they have a plausible and compelling understanding of the chosen location for their novel, in other words, they are "location-wise". A core element of this location-wisdom is to use telling detail to conjure a physical sense of place. The reader needs to be convinced that the story could emerge from this location, and thus the physical characteristics of that location need to align to the actions of the character. This conversation reminds us of the fragile nature of the special game involved in reading novels – the telling details of an author's crafted words need to incite the reader into creating a world with these words and their own corresponding geographies; however, if there is sufficient misalignment between these components, the "trick" will fail. There appears to be, therefore, a fine line between (successful) "disruption" and (ineffective) "jarring", demonstrating perhaps how easy it is for the magic of reading literature to go wrong, and conversely, how absurdly alchemical it can be when it goes right. Such fragility in reactions and relations are usefully summed by the ability of Lloyd Jones's *Mr Vogel* (2004) to produce polarizing responses from different reading group participants:

R2: Well I said something before and I'll say it again. If *Mr Vogel* is supposed to be a tour of Wales then I'm a banana sandwich.

(Survey Respondent)

F1: I didn't get it at all . . . I've never I think read such a disorientated, disjointed novel . . . It was annoyingly anarchistic in its structure . . . .And if that was supposed to be depth, it wasn't.

(Reading Group Participant)

R5: I liked it, it's kind of spacey.

(Survey Respondent)

R7: I absolutely love [My Vogel]. It's very difficult to get into at first [but] as soon as you go to Part Two, it's so well written and it's just, the whole book explodes in your head – it's just superb.

(Reading Group Participant)

Escaping through plotted reading offers new variations on person–place relations, perhaps offering new person–plot relations, with the act of plotted reading "dragging" (after Jones, Chapter 8) down the modern walls of categorization between the "real" and "imagined". A creation of a read-reality comes to define the reader, a version of the world that does not confuse the novel with a documentary source but embraces the ontological co-equivalence of plot and corresponding geographies. As a consequence, and in line with the potential identified in Chapter 7, literature changes its status from simply "make-believe" to become a generative device which can alter person–place relations and change how readers may understand their relations to the world. With this insight from the *Literary Atlas* reading groups, this book now progresses to explore how these groups, alongside their academic peers, engaged more broadly with the resources produced by the *Literary Atlas* team; Chapter 10 explores how these readers plotted their way through the distant and deep maps for the project's selected novels.

Notes

1 As Whiteley and Peplow suggest a "socio-cognitive approach to reading group" would identify that "two broad approaches to book groups have emerged: interactional research focussing on the discursive practices that these groups typically engage in (e.g. Benwell, 2012), and stylistic research looking at the ways in which the linguistic structures of literary texts relate to readers' interpretations (e. g. Peplow et al., 2016)" (2021:120). As a working alternative, Childress and Friedkin "advance an empirical and formal analysis of the cultural reception of texts . . . in which the intersection of social structure, individual readings, and interactive group processes all may enter into readers' interpretations of a novel" (2012:46).

2 Reading Groups questions for Perry's *The Essex Serpent* (2017:429) asked questions focusing around language choice and dramatic development, for example,

1. "I'll fill your wounds with gold", Micheal says. He means literally that he will make sure Cora is financially comfortable during their marriage in exchange for the pleasure of hurting her, but also that he will remake her as something more beautiful and interesting than she was before. Cora survived her horrible marriage, but was definitely damaged by it. What do you think the seams of gold are in Cora's character?
2. Many comparisons have been drawn between Sarah Perry's writing and the Victorian novelists who were writing at the time the book was set, including Charles Dickens and Wilkie Collins. Do you think this book feels Victorian, or contemporary?

3 The "Chapter 10" description which the reading group participant alludes to is this one:

> He was sitting above the Kestrel, looking down across the valley. It was strange to be up there alone, with the valley so quiet. The people he had lived with, the voices he had listened to, were all there under his eyes, in the valley. But all he could see from this height were the fields and orchards; the houses white under the sun; the grey farm buildings; the occasional train, very small under its plume of smoke. He had come up past Parry's farm and the sheepfold, and over the long climbing ridge of Brynllwyd. At first the paths had been wide, through the dense green bracken, past the last cottages and the occasional pool. Then, where the bracken dropped back, the wilder country of heather and whin stretched ahead. The paths became narrower, the air colder, and the mountains around took on different shapes, their moulding clearer. The sheep, in this great open stretch, were few and scattered. He could see the wild mountain ponies, but always in the distance. In the great silence of the mountain the song of the few birds, the strange high sound of the wind, came in a different dimension. At last he turned back from the open mountain, and crossed above the Kestrel to where he could see the valley again.
>
> (Williams, 2006 edition:283)

4 Echoing this comment, the following respondents remarked upon Dafydd's "location wisdom" to conjure up an affective sense of Bardsey/Enlli in words, and the consequence of this was the durability of this "read-reality" to remain with the reader as their definitive version of the island:

> I really enjoyed it for similar reasons. And thought it was, it definitely gave me a really strong sense of place, and it was quite a, you know, it was that inward-looking island, where you know you've very much felt they were trapped there, because of the waters, and they can get off and everything, and yet you still had a real sense of those huge skies and all the sea around you, so a real sense of openness in one sense, but also inward-looking and not able to escape, which was I thought really interesting, and really made me want to go and visit Bardsey Island. Which I didn't really know anything about at all. The island stayed with me.
>
> (Reading Group Participant)

5 Cyril's shop is introduced to the reader in *Sheepshagger* in the following excerpt:

> He just is, that's all, Roger replies. Then thinks for a moment and says, softer: -No, ee's alright, really, and will say no more until they enter Mad Cyril's shop, dark grotto festooned with hanging T-shirts and lengths of material all colours and a thousand eyes of watchfaces, clockfaces and the myriad glint of accessories, rings and necklaces and lighters and buckles for belts and clips both money and roach pewter and nickel and silver. Draped bootlaces like worms and pot-pipes some smooth stone, some wood elaborately carved, and bottles of poppers

arrayed in trays upright like attentive snails and small sculpted dragons and warlocks on glass shelves, shallow bowls patterned with Celtic knots and woolly hats and socks and underwear all patterned green and white with the dragon emblazoned red and rampant.

(Griffiths, 2001:175)

6 One extract that exemplifies the nature of Gallie's community, which corresponded closely with the reading group's own familial experiences describes Gwen's funeral:

> Gwen's funeral that afternoon was one of the largest remembered in Cilhendre. The numbers were swollen by the morbidly curious, but most of those present went to show their deep feeling for the mother, a little, punch-drunk, yellow widow. Mrs. Nixon sent a wreath, Elwyn Jeffries sent a salmon to help feed the visiting relatives, Everynight brought his five shillings fox money because he claimed that Maggie's beer was flat that week, and Maggie herself sent down a dozen bottles of stout so that Mrs. Evans could keep up her strength. The emotion and the crying at that funeral would have more than satisfied Liz and Rita Manager. The little baby was buried too, after his inquest, his moment. A short formal service for the unbaptised, with policemen for pall-bearers, and no flowers, no tears, no hymns. But the ground was consecrated, all was well.
>
> It was a sadly moving day for Cilhendre. But, somehow, satisfying. Emotions were released and relaxed by the sublimity of Christian burial. For Gwen they could believe in it all. John Nixon, representing his family at the service, felt that what had been mumbo-jumbo for his father was for Gwen the necessary truth, and was furious with his own illogicality. He was moved almost beyond bearing.
>
> The congregation was uplifted by the service and the singing. Some inner need was satisfied, the opium of the poor was a powerful drug, and they left the chapel feeling refreshed and belonging and reassured for the time.
>
> (Gallie, 2011 edition:157)

7 One example that displays this aspect of Griffith's crafted words is given as follows:

> Through the Penparcau estate the bus moves and out again of this warren, passing the huge billboards there from the large banks designed to attract the potential student custom arriving from southern England; the boons on offer, a huge £20 note bigger than some of the rooms Ianto has slept in, bigger than the living rooms of the houses on this estate. Out on to the mountain road and Ianto sees the small huddled villages and seeping sheep and the grey-bellied sky over it all and the meandering river the colour of strong coffee and the vast expanse of Borth bog, steam rising, it stretching to the blurred horizon over the far estuary, where the mountains swell above the broken white line of Aberdyfi, a weak shaft of sunlight spilling across their grey and distant crests. Machynlleth is a small flurry of opening shops and busy workers and wet traffic seen through rain-slick glass; headlights burning through the drizzle and the streetlights still bowed and blushing, somehow dreamlike and unreal as if it is the hours of night yet, but the sky has not darkened nor will it. Ianto is carried over the swollen river straining its banks and he sees mountains rising rising, small whitewashed cottages chimney-smoking and mist and more green mountains rising, these parts of his own small country ancient and arcane this hard demanding land he was born on and into its small roads and sheeptracks leading maybe to marvel and secrecies revealed. Umbilicus unseen never to be snapped the brutal beauty of this place has battered itself into his blood, his brains.
>
> (Griffiths, 2001:195)

8 Some participants remarked, for example, that in latter stages, *The Rebecca Rioter*'s dramatic arc deviated from the geographies it had painstakingly created:

F9: . . . the second half was very contrived I mean, you know. Why would you leave a gun with your name behind? Why would you go back [to the village] – you know? What about your sense of self-preservation?
M1: Yeah. The plot seemed a bit [trails off]. (Reading Group Participants)
F2: You know, there was Amy Dillwyn's one, and then she did the absolute horror of horrors, she brought in a Dickensian coincidence [laughs] to make it sort of . . . I don't know, more interesting, I suppose, which it had nothing whatsoever to do with the story or the Rebecca rioters or anything else. She made one of the characters turn out to be the long-lost lord or somewhere or other and I though 'oh, please'.
F3: Yes, that's exactly the reaction, isn't it. Give it a rest! [laughs].
(Reading Group Participants)

A different audience closer to the time of writing may have forgiven these dramatic conceits given the nature of their own corresponding geographies and the expectations of a novel at the time, but for these reading groups, their use took them out of the novel and magic of the crafted word failed.

10 Engaging publics

The *Literary Atlas* online

As we have seen, *Literary Atlas* adopted a plotted reading approach to engage with, and subsequently map, 12 novels written in English and located in Wales. From this emphasis, distanced and deep maps were produced, emergent from the authorial locations and corresponding geographies identified by the *Literary Atlas* team. As Chapter 8 introduced, as part of the project, we were committed to a series of evaluation works of these curated resources (see also Chapters 11 and 12). As part of this process, the *Literary Atlas* team wanted to gain feedback from voices both within and beyond the academy about the approach of the project and the variety of maps curated through it. To that end, we invited a set of peer reviewers and participants from our 21 reading groups (see Chapter 9) to engage with the distant, deep, and plotline maps produced on the website. We asked these individuals to comment on the accuracy and usability of these resources, and also insights they generated. As this chapter will outline, all reviewers considered the *Literary Atlas* website to be an effective and instructive resource, with comments including positive engagement with coverage and useability. In terms of mapping materials (including the deep mapping and plotline components), it was considered that these offered interesting and additional perspectives to not only understand the novels selected but also the wider geographies they corresponded to. As a consequence, engagement with the *Literary Atlas* materials heightened curiosity and interest in both Welsh writing in English, but also Welsh places; interaction with these plotted materials appeared to generate a strong sense of connection to literature and geography beyond that mobilized by a passive reading of the novels in isolation. This engagement suggested therefore that the plotted approach, and the *Literary Atlas* materials emergent through this, had the potential to generate interest in local history, community, and literary tourism more generally (see Chapter 11).

Peer review and reading group evaluation of *Literary Atlas* online resource

Twelve peer reviewers were invited to review the *Literary Atlas* materials, reflecting the diversity of subject areas we felt were influenced by and interested in the confluence of intellectual currents sweeping across the social sciences, arts, and humanities. These reviewers included experts on contemporary Welsh literature, historical Welsh fiction, critical studies, fantasy fiction, folklore, horror, history,

DOI: 10.4324/9781003321248-10

politics, culture, creative writing, and the publishing industry in Wales. Allied to these academic and industry voices, participants from the *Literary Atlas* reading groups were invited to engage with the project materials online (and from this call, 46 book group participants responded). It was agreed that responses from both experts and the broader public would be anonymized.

As Chapter 4 outlined, the *Literary Atlas* team undertook a reflective process to shortlist novels, and through this evaluation exercise, there was a further opportunity to gain feedback on how the general public as well as academic peers responded to the project team's choices. In general terms, the *Literary Atlas* materials were considered by both sets of respondents to be well set out, clear, and informative, as the following statements suggest:

> I think it is excellent, a great idea, well-executed.
>
> (Reading Group Participant)

> The range of scholarship should be commended and the website is sophisticated in its approach, yet comprehensible and accessible without compromising the depth of its analysis. *Literary Atlas* is a valuable, exciting addition to the study of Welsh literature.
>
> (Peer Reviewer)

> A very interesting project! Thank you for introducing me to material I otherwise would not have read.
>
> (Reading Group Participant)

> *Literary Atlas* is a useful and interesting project which is imaginatively and attractively presented.
>
> (Peer Reviewer)

> The information on the website is factually accurate and links to outside sources which offer contextual information, which will be useful for the general reader. Broadly, the information is clear, comprehensible and insightful. It is accessible and usefully complemented by images, extracts, maps and audio clips.
>
> (Peer Reviewer)

More specifically, respondents commented with regard to how the selected shortlist of novels deviated from the generally established canon of Welsh writers (but notably perhaps, respondents did not directly comment on the desirability of this); some applauded the focus on Welsh writing in English, whilst some respondents identified specific authors they felt should have been included:

> The novels chosen as its focal point are a mixture of historical and contemporary works, some 'classic' but others less well-known, including a few which I had not previously been aware of.
>
> (Peer Reviewer)

> Anglophone writing by Welsh authors can occasionally fall into a no-man's land, where it is not regarded inside Wales as being 'properly' Welsh, but is also neglected or overlooked in the world outside Wales as being 'too Welsh' to be of general interest. A dedicated resource of this kind is therefore useful. The site should make a valuable contribution to the discovery and study of Welsh writing in English.
>
> (Peer Reviewer)

> Please add more authors and works. Wales is quite underestimated, literally speaking, at least outside Britain.
>
> (Reading Group Participant)

> Really enjoyed visiting. Thank you. It's a valuable resource, but sense you need some more 'familiar' books in addition to the ones selected.
>
> (Reading Group Participant)

> Please include the works of Emyr Humphreys and Dannie Abse.
>
> (Reading Group Participant)

As the earlier responses suggest, some of the novels chosen may have been not widely known or appreciated, perhaps because they were too "literary" in nature and therefore beyond the awareness of many in and around the reading classes, or perhaps because an online platform to celebrate the diversity of English-language novels set in Wales has not been, heretofore at least, widely available. Comments of this type could be seen to resonate with the reading group survey responses noted in Chapter 9; here, when asked about their favourite (or most well-known) Welsh writers in English, Dylan Thomas was returned as the number one choice. As Thomas predominantly crafted his plots through plays and poetry, he was not long-listed for this project, but his popularity nevertheless draws attention not simply to his acumen but also perhaps the relatively small pool of "known" Welsh writers in English (especially when compared to the quality of authors in this category[1]). Further comments were made regarding the absence of self-published works on the short list given the growing popularity of these in recent years (as one Reading Group Participant noted, the website could: "Add more authors, especially those published outside Wales or self-published with a connection to Wales") but nevertheless it was broadly acknowledged that

> The parameters set for the chosen content seem sensible, with a valid statement of why the site's focus on Anglophone writing establishes the site as a useful starting point for extension into Welsh-language writing or other media.
>
> (Peer Reviewer)

Further comments were offered about the coverage of the short list. As the following responses indicate, these referred both to geography and also to specific classifications of fiction (e.g. young adult, or children's literature) which, in hindsight, offer intriguing possibilities for further iterations of the *Literary Atlas*:

> In terms of gender, the choice of authors is also well-balanced, and the novel's geographical spread balances the south with the frequently neglected north of the country.
>
> (Peer Reviewer)

> Please add something to represent West Wales, but on the whole an impressive achievement.
>
> (Reading Group Participant)

> Could you put children's books on there. Picture books, middle grade and young adult as there are a lot of those.
>
> (Reading Group Participant)

In terms of usability, in general, reviewers "greatly enjoyed" (Reading Group Participant) interacting with the maps in the *Literary Atlas* resource:

> the map on the homepage is great, as well as how clear the layout/design is.
>
> (Reading Group Participant)

> The pages are clear and easy to navigate.
>
> (Reading Group Participant)

> In terms of usability, I found the site intuitive and easy to navigate, as well as attractively laid out.
>
> (Peer Reviewer)

For some, the interactivity of the website took a small time to master:

> It took a lot of scrolling to find out what the website was about but after that I really enjoyed it.
>
> (Reading Group Participant)

Reflecting the nature of many websites and the differential paginations produced when displayed on varying technological hardware or screen size, a number of respondents found it "a little trickier to use [the *Literary Atlas*] on a smartphone" (Reading Group Participant).

It is important to note that the *Literary Atlas* resource was designed for interaction on all screen sizes and technologies, but is perhaps best appreciated on laptops, tablets, or computer screens (rather than cell phones). Small screen use is possible, and was directly designed into the resource as, as Chapter 11 will outline, it was hoped that it would provide the basis for in-field interaction through literary tours.

Navigating maps

Key comments were also focused on the variety of maps curated by the *Literary Atlas*. These emerged from the catalogue of Welsh writing in English generated from the archives of Cardiff University, Swansea University, and the National Library of Wales (see Chapter 4). The catalogue itself was deemed to be "particularly useful" (Peer Reviewer) by many:

> both as a historical database of Welsh writing in English which should provide assistance to researchers or writers on historical and contemporary Wales, and as simply an efficient resource for readers interested in this area to discover further reading material.
>
> (Peer Reviewer)

> the 'library' section of the website is also a very welcome feature, which provides a fuller map of Welsh literature in English.
>
> (Peer Reviewer)

As Chapter 4 outlined, it was from this database that distribution maps of the novel's dominant locations were generated, and users commented that these,

> 'Coverage' pages, showing the spatial dimensions of the novels' content, were a striking and efficient way of visually displaying this information.
>
> (Peer Reviewer)

Coupled to this,

> It was interesting to see the graphics depicting the areas each book covered.
>
> (Reading Group Participant)

From this basis, the shortlisted novels were chosen, and deep maps were generated directly from a plotted reading of these novels. The multimedia layered nature of these deep maps and "combination of pictures & maps, quotes, opportunities to listen, alongside explanations" (Reading Group Participant) "added interest" (Reading Group Participant) and were considered a useful innovation by our reviewers:

> Firstly, I think that the mapping visualisations are working really well. I really like the level of detail that emerges as you move through the layers or deep maps.
>
> (Peer Reviewer)

> *Literary Atlas* is an impressive resource and its mixture of maps, extracts, interview excerpts, images and audio-clips usefully complement the texts currently detailed.
>
> (Peer Reviewer)

> I liked the interspersing of comment and quotes with actual images from the given locations. The audio recordings were good, though the quality was variable.
>
> (Reading Group Participant)[2]

> They are a brilliant combination of map and essay. The essay element not only summarises the [dramatic] plot, but it also supplies information on the places and gives inspiration on interpretation. The map physically locates the plot and provides additional information to specific places.
>
> (Peer Reviewer)

> The inclusion of maps really brings the novel's setting (and the setting of the original tale) to life.
>
> (Reading Group Participant)

Visualizing plot-points cartographically was therefore universally seen as a useful addition to the repertoire of literary geographies. Although the quality of the supplementary audio used was dependent upon the quality of microphone technologies and sound-proof studios (and reflecting the expansion of skills and methods now required by the literary geographies will become a fundamental criterion to be factored into future project planning), in aggregate, these deep maps were deemed to be an aid to understanding, as the following respondent put it:

> It was just great to see plotpoints on an actual map instead of trying to construct it in my mind's eye – I am no good at visualising things in space.
>
> (Reading Group Participant)

From these deep maps of plot-points, which set up the wider context of the novels and their corresponding geographies (see Chapter 6), particular plotlines were then curated through this expanded territory. These "-lines" added new routes through the plot, and the digitized cartographies gave the user virtual mobility between plot-points, augmented by the deep mapping materials and explanations at each stop on the route. Respondents enjoyed these additions, commenting both on the interactivity and movement of the maps as each chosen plot-point was visited:

> I like the lines on the map and explanations to what happens . . . [for example] like moves to Vancouver etc.
>
> (Reading Group Participant)

> I loved the plotlines, seeing a map of familiar places and imagining the characters moving along it.
>
> (Reading Group Participant)

For many, these plotlines and their associated deep maps combined effectively:

> The plotlines added new visual dimensions and encouraged better awareness of the place's essay.
>
> (Peer Reviewer)

> The plotlines sections work very well – lots of information with useful illustrations which builds up a really rich sense of context for the novel.
>
> (Peer Reviewer)

It was particularly intriguing that some respondents commented on the utility of these plotlines for novels that were less familiar to them:

> More generally, I very much enjoyed the material on some of the novels which I don't know as well, particularly *The Owl Service* and *Twenty Thousand Saints*. That suggests that for a general reader this is coherent, engaging and informative.
>
> (Peer Reviewer)

Indeed, many respondents focused specifically on the "enjoyment" (Reading Group Participant) they received from the "new" and "fresh ideas" (following Bradbury, 2008:108, see also Chapters 2, 7, and 13) they received through engaging with specific plotlines:

> I enjoyed that the plotlines showed me some interpretational ideas I had not thought of before reading them.
>
> (Reading Group Participant)

with each respondent picking out their favourite:

> I think the Lloyd Jones plotline works really well, and the points he makes about his difficult negotiations of personal freedom are fascinating. The photographs are really touching. They would add a whole other fact/fiction conundrum had they been included in the novel.
>
> (Peer Reviewer)

> In relation to *Pigeon*, the overall assessment of the text's presentation of space is strong, varied and suggestive. The combined use of photographs, maps, interview excerpts and extracts from *Pigeon* works very well in the 'plotline' part of the page. Readings of the novel by Alys Conran are a particularly good touch . . . [as is] the map function which demonstrates the route Pigeon follows from his home to the juvenile centre . . . [it] help[s] visualise the distance and direction of travel characters follow.
>
> (Peer Reviewer)

> Breaking the longer analysis of *The Owl Service* into four parts to aid readability was a good idea.
>
> (Peer Reviewer)

> [Revenant] is a novel which lends itself very well to literary geographical mapping as it [appears] so embedded in place and effectively reads as a walk from one place to another.
>
> (Peer Reviewer)

For many, these plotlines helped to offer insight and understanding into the power of plot and a plotted reading approach to literature, as the following respondents comment in relation to Mr Vogel:

> Firstly, I think that the mapping visualisations are working really well. I really like the level of detail that emerges as you move through the layers or deep maps, but also the panning out so that the reader can envisage the enormity of Jones's walk around Wales. This is totally at odds with what is an incredibly abstract and impressionistic text. I don't know how you have done it but is strangely satisfying to have it detangled like this!
>
> (Peer Reviewer)

> I found the info on Mr Vogel more interesting than reading the book. It clarified some details and gave fascinating insights into Mr. Vogel and its author in particular.
>
> (Peer Reviewer)

> The plot descriptions helped to understand the book.
>
> (Reading Group Participant)

> It did not change my perception. Instead, it confirmed the theories I had already developed. That identity is very much bound to place, for example.
>
> (Reading Group Participant)

For others, the combination of deep mapping and plotlines provided accessibility to and awareness of broader themes which lay within novels; for example, with respect to *The Rebecca Rioter*, these maps made one respondent "aware of the transgender theme and queer romance" (Reading Group Participant), and another become aware of "the lesbian relationship which changed my reading of the book" (Reading Group Participant). These plotted reading products therefore were impactful in changing users' understanding and overall relation to the novels, calling forth a range of important themes. Perhaps most obviously, however, these products emphasized the significance of cultural and physical geographies of plot, as the following respondents make clear:

> These resources helped me understand the background of the story better.
>
> (Reading Group Participant)

> They deepened my understanding . . . added depth and specificity and they put an interesting historic aspect to them.
>
> (Reading Group Participant)

> They make you think about the locations more.
>
> (Reading Group Participant)

> They emphasised the location's centrality to the story.
>
> (Reading Group Participant)

Conclusion: the plotted approach generating interest in Wales and its literature

We have seen throughout this book that a plotted reading can help to converge once separate ontologies, in particular, what the modern constitution categorizes as the "real" and "imagined", into one co-equal plane. It can be seen from responses to the *Literary Atlas* website that the curated maps shared here can reinforce this process, making the literary geographies of these novels "more real", rather than simply "make believe". In general, these maps helped users "think more about the interactions between text and world" (Reading Group Participant), but more precisely, they "gave the stories much more depth", and "made them more real" (Reading Group Participants). Visualization through digital cartography thus appeared to affirm the "reality of the read" to these website users, conferring to it a stronger sense of what the modern terms "real". The geographies which were seen to correspond to the crafted words of the author were, through these cartographies, seen to be more strongly sutured to them, and together the impression was given that these read-realities could and should be taken more seriously, or at least as seriously as that which wholly conforms to the categories of the modern constitution. Allied to the insights from Chapter 9, we can see here that plotted reading can be framed again not simply as an escape from existing person–place relations (however important that insight is) but potentially as productive of new worlds in which corresponding geographies converge with stories (and vice versa).

From the responses to the *Literary Atlas* website, it can be suggested that these combined insights not simply have a philosophical, but also a practical consequence. Through interacting with these combined curated maps, respondents suggested they were encouraged, and perhaps even compelled to find out more about plots – be they about the novels in question, their corresponding geographies specifically (in a local or regional sense), or more generally (about Wales, its geography, society, and literature). The plotted approach, therefore, can be seen to be generative of interest about Wales, its culture, and its literature, as the following comments suggest:

> It gave me more of a sense of the histories of particular places – very helpful way in to history, which is something I find hard/overwhelming in general.
>
> (Reading Group Participant)

> It made [Wales seem] more regionally different; it is not so much a single space but a collection of spaces.
>
> (Reading Group Participant)

> It added the border country as a literary and socio-cultural space for me. Before, to me, Wales was defined by exploiting natural resources and an inherent feeling of communal and national belonging, produced and contrasted by England as the opposite.
>
> (Reading Group Participant)

> It gave me a different view on the country and created a certain connection to it.
>
> (Reading Group Participant)

> It showed me that there's so much more to read! Very exciting!
>
> (Reading Group Participant)

> I felt it might be interesting to visit some of these places.
>
> (Reading Group Participant)

Evaluations of the *Literary Atlas* website suggest therefore that the plotted approach can generate interest in plots, and their corresponding geographies. In the same way, that plotted reading can be understood as a form of escape into the borderlands of Welsh writing in English, sharing a variety of creative maps of these plots had the effect of encouraging *Literary Atlas* users to find out more about these corresponding geographies, in all of their forms. The plotted approach therefore demonstrated the potential to not simply encourage interest in and identification with locality and (Welsh) society but also Welsh places through literary tourism. To explore these possibilities further, the *Literary Atlas* team invited reading groups and the general public to join in "plotted excursions" through key plot-points and corresponding geographies for each of the 12 selected novels. This facet of literary geographies we will explore in more detail in Chapter 11.

Notes

1 In turn, this perhaps reflects the absence of key writers in this bracket on the school curricula in both Wales and the UK more widely, see Chapter 13.

2 As a consequence of this feedback, the audio files were upgraded wherever possible on the public version of the website.

11 Engaging publics

Creative excursions into the borderlands

Introduction: plotted reading as exploration

As we have seen, the *Literary Atlas* was engaged with a plotted approach to literature, identifying and recognizing not simply how the geographies and drama of an author's crafted words are constitutive of one another but also how these plots come to correspond with other locations through the practice of plotted reading. From this platform, the *Literary Atlas* team identified a range of plot-points from our selected novels, and created distant and deep maps which were facilitated by and disseminated on the digital resource (www.literaryatlwas.wales). Trajectories through these plot-points and their corresponding geographies – or plotlines – were then curated, offering a route through the broader context of the selected novels and new perspectives from which to (dis)orient the reader to the plot as a whole.

As we have seen in previous chapters, the *Literary Atlas* team invited reading groups to engage with these combined materials. Through so doing, further insight was gained into how the practice of reading can be seen as geographical in nature (see also Hones, 2014), but more specifically, as a form of escape from existing person–place relations and an often enjoyable disorientation into new assemblies of characters, drama, and locations which are inevitably tied into the reader's own. In this sense, the practice of plotted reading can be understood as a form of exploration, a journey of disorientation into a new read-reality. Sharing the insight experienced by Roald Dahl's Mathilda, through plotted reading, it was possible to be "transported . . . into new worlds" and be introduced to "amazing people who lived exciting lives" (Dahl, 1988:15); through reading, we can "travel . . . all over the world while sitting in [our] little room" (Dahl, 1988). As we have argued in the preceding chapters, for *Literary Atlas*, this is a significant acknowledgement. Recognizing (plotted) reading as a form of exploration is part of the wider political project to dismantle the imposed modern categorizations of "real" and "imagined"; affirming that a read-reality is not to be ignored as simply figurative or peripheral to the modern realities of life, but rather recognized as a co-constituting ontological realm which helps readers to orient themselves in the world and give them a sense of belonging within it. In other words, plotted reading helps us locate, and build, our own borderland.

DOI: 10.4324/9781003321248-11

As Chapter 1 stated, author Phillip Pullman describes the borderland as the geography,

> that opens up between the private mind of the reader and the book they're reading. It'll be different for every individual, because while parts of the borderland belong to the book, other parts belong only to that particular reader – to us: our memories, the associations we have with this or that particular word or landscape, the aspects that resonate with our own individual temperament; so whereas many readers might be reading the same book, no two of them will read it in exactly the same way. However, we can come back and talk about it; we can tell others about our experiences of it, and compare our part of the borderland with other people's.
>
> (2017:259)

From the perspective of this book, the plotted reader therefore creates a borderland between their own lives and the crafted words of the author. More specifically, the borderland is a literary geography which emerges from the coming together of a plotted reader and the affinities they forge between the author's crafted words and a range of corresponding geographies which together come to co-constitute their lives. In one sense, the borderland could be understood as the "property" of the reader (in the sense that they own it, in Pullman's words earlier: it "belongs to us"); but in the spirit of this book, it has greater significance than simply a space over which ownership can be claimed. From the perspective of plot, the borderland forms a new read-reality which co-constitutes the person–place relations of the reader. The borderland not only is the place where they live but also is a place which defines who they are.

As part of the evaluation of the project, the *Literary Atlas* team wanted to explore this new literary geographical space in all its facets and dimensions (including the plotted readings of the novels, the perspectives of the authors through interviews, the plotlines produced by the *Literary Atlas* team, and the key corresponding geographies produced). We were particularly interested in how some of the potential corresponding geographies for plot-points may influence our reader's understanding of and relations with these borderlands. To that end, we undertook what can be termed as "plotted excursions" into these borderlands – trips which explicitly sought to bring together key plot-points and their corresponding geographies from the website, with their associated physical places in person. We invited the *Literary Atlas* reading groups to participate in these plotted excursions, as well as inviting other interested publics through the project's social media platforms. In the spirit of working through what these relations may "do" (see Chapter 3, and Anderson, 2018), we were intrigued how a physical encounter with these plot-points and some of their corresponding geographies (at least from the perspective of the *Literary Atlas*) may affect or otherwise mobilize participants; in effect, we wished to explore what new insights into the assemblage of novels, maps, and localities may be produced as a consequence of these encounters?

The chapter evaluates these plotted excursions. It begins by outlining conventional approaches to literary excursions – most commonly understood through recourse to theories and practices of tourism. The chapter will argue that a literary tourist predominantly engages with a physical place, often through the passive consumption of artefactual (see Chapter 3) pre-packaged commodities. The chapter will suggest that plotted excursions offer a significantly different mode of literary tourism when compared to this mainstream. Influenced by the relational sea change and associated currents of culture, space, cartography, and creativity (see Chapter 3), "plotted excursions" can be understood as a new form of "creative tourism" (see, for example, Richards and Wilson, 2006; Richards, 2011). The chapter will outline the nature of the "plotted excursions" run by *Literary Atlas*, involving the coming together of plotted readings, website materials, and corresponding geographies, alongside the opportunity (as Pullman suggested earlier) for participants to "talk about" their borderlands, and "compare our part of the borderland with other people's" (2017:259). This chapter begins with an evaluation of these plotted excursions into the borderland through contextualizing these practices with regard to conventional literary tourism.

Conventional literary tourism

Conventional literary tourism can be generally understood as "tourism associated with 'places celebrated for literary depictions and/or connections with literary figures'" (Squire, 1996:119, cited in Hoppen et al., 2014:2). Such tourism is a long-standing phenomenon (see Müller, 2006:215), with Yiannakis and Davies identifying its initial development in "the eighteenth and nineteenth centuries" (2012:34). It is a practice that has become a "commercially significant phenomenon" (following Watson in Hoppen et al. (2014:3), which is particularly prevalent in anglophone territories. Hoppen et al. (2014) go on to cite Watson who suggests:

> literary tourism has become so 'naturalised as a cultural phenomenon in the British Isles [for example] that one sees literary sites detailed in guidebooks and marked on the road map, and expects (and feels expected) to visit the museum shop and to buy the soap, the postcard and the bookmark'.
>
> (Watson, 2006:5, in Hoppen et al., 2014:37)

As Watson suggests, the development of literary tourism in the UK indicates the importance of culture, and specifically literature, to those who engage with them; in particular, their willingness to translate this interest into an investment of their time and money into satisfying a broader engagement with literary geographies.[1] However, it is also notable that, as Watson implied earlier, the "naturalization" of literary tourism is also in practice the development of a particular mode of engagement with literary geographies. This is a phenomenon that involves a key tourist stakeholder (e.g. a unitary authority, a quasi-governmental tourism enterprise, or a private entrepreneur) formally associating a specific location with a literary figure (e.g. an author, as Squires noted earlier, and we will expand upon below). In this way, literary tourism is packaged in a "heritage tourism" mode (see, for example,

Herbert, 2001), treating authors in the same way as (other) key historical figures (e.g. military leaders, members of the aristocracy or monarchy, inventors, or other culturally significant individuals). This "naturalized" mode of tourism packages the author's life into specific locations which can be scripted, curated, or otherwise produced to transfer quasi-educational facts to tourists (including, for example, the information about the birthplaces of writers, the homes in which they lived, or significant cultural sites in their lives (see, for example, the "writer's plaque" scheme curated by Literature Wales and available to view on the *Literary Atlas* website, see www.literaryatlas.wales/en/maps/plaques/)). In this way, this version of literary tourism may not require the visitor to be directly engaged with the books of an author (in the same way that visitors to a stately home may not be directly motivated by a particular member of the aristocracy who lived there); rather tourists are broadly interested in gaining a general knowledge of the cultural history of a region that are visited, and literary figures have been included in this diverse roll call.

Acknowledging the inclusion of literary tourism as part of a broader mode of heritage tourism is important. In concentrating attention (to some extent at least) on the lives of writers (for example), it focuses on the ways in which literature is treated like any other key industrial or cultural artefact; an entity which can (firstly) be humanized through the life of the author; secondly, can be spatialized by tying it into the history and geography of a region; and then (thirdly) be commodified by requiring a monetary value for a personal encounter with the location or its associated information.

More intriguingly (and as Squire acknowledged earlier), this naturalized version of literary tourism is not always or wholly focused on the producers of a literary artefact (i.e. on authors). In some cases, an author's fictional characters can be given the heritage tourism treatment (for example, Conan Doyle's fictional detective Sherlock Holmes has a heritage marker on the site where he first met colleague Dr John Watson, see The Londonist, 2022). By implication, therefore, this version of literary tourism is also extended by focusing on the places sited "within the story" (after Mansfield, 2015:2). Such places occur when, in Mansfield's words, authors formally adopt "topographical accuracy . . . to site their purely fictional events" (2015:33). In a broadly similar vein to the distribution maps produced by *Literary Atlas* from both the geoparsing exercise of the library catalogues (see Chapter 4) and the subsequent regionalization of plot-points for the selected novels (see Chapter 5), such alignment between locations within a story and (what the modern constitution terms) the real world produce districts associated with authors that, from the perspective of this form of heritage tourism at least, appear to "hover above" the real (to use the words of Piatti and Hurni, 2011:218). According to Mansfield, such districts may include

> 'Catherine Cookson County' in Northumberland (Herbert, 2001), . . . 'the Brontes' Yorkshire', 'Hardy's Wessex' (Squire, 1994), [and] 'Wordsworth's Lake District' . . . (Watson, 2006).
>
> (Mansfield, 2015:3)

This expansion of literary tourism in effect treats characters and fictions in the same way as it treats the lives and locations of authors. In this mode, a heritage tourism approach "serially reproduces" (after Richards and Wilson, 2006) literary "place-packages" irrespective of whether the unit of artefactualization is a character, author, or location (a point we will return to below). In this mode, tour guides, brochures, and other information are produced which "direct[s] expectations, influence[s] perceptions, and thereby provide[s] a preconceived landscape for the tourist to 'discover'" (after Weightman, 1987:230, but also cited in Pritchard and Morgan, 2001:167). In this mode, brochures, guides, and cartography offer metaphorical "keys" or "passwords" to understand a definitive truth behind the places of writing or setting. Through doing so, they produce not only a literary geography which is fixed and durable in the modern tradition but also a tourism practice that is commoditized, scalable, and relatively passive in terms of the experience of the tourists themselves.

To some degree, this commodification of author, character, or location (irrespective of its modern ontological classification) is a significant move: here characters and their literary geographies are not marginalized or ignored as simply "make believe", but recognized as important to the lives of tourists who wish "to correlate fictional locations with some markers of reality" (after Robinson, 2002:40). As this occurs, the possibility is raised for tourists to begin to enjoy the confluence of modern categories in practice, and through its capacity to generate revenue, the power of literature is acknowledged by a new entrepreneurial audience. However, it is important to note that such "one size fits all" commodification does not seek to harness the power of reading, nor does it begin to recognize the co-ingredience of drama and location as undertaken by the plotted approach. As such, it does not go so far as to actively subvert or remove the separation between the modern categories of "imagined" and "real", or acknowledge the sizeable potential of its practice.

Thus, the naturalization of this mode of literary tourism strongly resonates with what Chapter 3 has identified as a sedentary approach to both place and literature. In this approach, texts and geographies are conceptualized as singular in nature, as entities (or artefacts) with fixed and durable meanings. However, as Chapter 3 has also identified, from a relational perspective, entities are better understood as processes, enjoying a multiplicity of meanings and interpretations as they come together with a range of other active components in broader assemblages. In such a relational framing, intrinsic to the plotted reading approach (for example), the practice of tourism becomes less about the passive acceptance of fixed narratives or histories provided by tourism producers, principally because such singular, authentic texts and places are no longer recognized. Rather tourism becomes understood as a practice of co-constituting place in the moment, a process of actively seeking out unique, emergent experiences which cannot be wholly programmed or predicted, but are rather co-produced by the coming together of discrete components which all have a degree of agency over the process and outcome. In short, literary tourism has become "phenomenal" in nature. From this perspective, engagement with literary geographies in practice is not limited by authorial intent or a marketeer's slogan, but rather emergent through the "complex production of meaning and effect . . . from the dynamic interaction" (after Drucker, 2008, cited in Barnes, 2013:166) between a reader, their corresponding geographies, physical locations, and the work of the author. Relational approaches thus counter any passive "reading" of literary tourism, and offer in its stead a creative, agential, co-production of literary places through practice. As Anderson and Stanley (2019) note, literary tours, in this view, are not pass*words* to open up definitive and singular readings of any literary geography but rather pass*ports* which invite the reader on a journey into an emergent world that they directly contribute to.

Tourism motivated by a relational approach to literature is thus fundamentally different in philosophy and practice to conventional literary tourism. Participants in this mode would not "discover" a passive commodity, or even "recover" (after Weir, 2002:199) something they've lost, but rather embellish, contextualize, and perhaps even rupture their own read-reality – their own borderland. It is this potential for literary tourism that the *Literary Atlas* project sought to explore and evaluate. The *Literary Atlas* project team invited reading groups and the general public (via open calls to public libraries and through *Literary Atlas* social media accounts) to "plotted excursions" through key plot-points and corresponding geographies for each of the 12 selected novels. In detail, each excursion was approximately half a day in duration, predominantly oriented on foot (although some trips involved a short bus ride to connect together a series of corresponding geographies),[2] and finished with a complimentary meal in a venue that enabled time for small group discussion. Participants carried micro- and smartphones on each excursion in order to record their encounters (individuals were also invited to take photos and make notes to record moments of interest and either post these to social media or use them as a prompt for later discussion). Audio recordings were also taken, then transcribed and thematically analyzed by the *Literary Atlas* team. All participants were encouraged to engage in dialogue and share their reactions to the corresponding geographies they encountered and align these with their own plotted reading of the novel in question. Attendance was therefore premised on having familiarity with the book which was the focus of the excursion, but beyond this, there were no "experts" formally designated, and the itinerary was loosely improvised around key plot-points. Plotlines were available through a beta version of the *Literary Atlas* online resource, whilst specific plot-points and associated media were also available through printed handouts.[3]

Plotted excursions into the borderland: "passports" to "fresh detail" (after Bradbury, 2008:108)

Experimenting with a plotted approach to literary tourism thus opened up possibilities for evaluating how the plot-points of a novel come together with an array of corresponding geographies through a walking encounter. As we will see below, participants in the *Literary Atlas*' plotted excursions suggested the experience of this new form of literary tourism was affective in nature: plotted excursions were an activity that mobilized the relations between book and reader and added an embodied dimension to the established assembly of plotted reading. In short, plotted excursions were an enjoyable, embodied experience, as the following participant usefully sums up:

> it made me relive the books in a way I never would have done otherwise, it brought a lot of details back to mind and showed me new details.
>
> (Excursion Participant)

This process of "reliving" the plotted readings of each novel thus appeared to draw a new level of awareness and appreciation of not simply the value of a plotted reading – with the participants refreshing their knowledge which was prompted by their specific engagement with the novel, but also adding a fresh contemplation of the importance of the borderlands they had conjured through this process. Coupled to these components, the experience gave them a new appreciation of the skill involved in crafting an effective piece of literature from which this world-building was possible. As the following participants stated:

> It made the book more memorable having visited the location.
>
> (Excursion Participant)

> The tour made me more intrigued with the book and its plot.
>
> (Excursion Participant)

To enable these insights, participants remarked on the need for active involvement, consideration, and reflection on their behalf. Individuals were aware that as they walked through plot-points they were not simply being given a package of pre-established and fixed artefacts, but were rather actively retracing the steps of a character in a borderland world that they had already conjured through their reading. Such a stimulating and perhaps even uncanny experience was articulated by participants in the following ways:

> The [trip] added to the adventure of [reading] in a way, and made me part of the journey in more obvious ways. I had to retrace my steps, as it were, to find what the characters already knew was there.
>
> (Excursion Participant)

> I enjoyed it very much indeed. I felt like I was following my own road map.
>
> (Excursion Participant)

The following participant on *The Owl Service* plotted excursion extends this reflection by commenting upon the embodied and sensorial nature of their experience:

> it was wonderful to see the 'actual' location – the quiet, just the cries of buzzards and the stream running over the rocks. Driving up to Bwlch y Groes brought to life a sense of danger from the twisting bends and steep drops on

which the motorbike in the story crashed. The ranges of mountains in view from the top was breathtaking, and the distances brought home the sense of isolation from the rest of the world. The farmhouse was hard to see from the road which made it seem all the more shut off and inward-looking and claustrophobic. Meeting and chatting to a couple of local people offset that sense of loneliness by their friendliness – in a small place, everyone speaks to everyone else. The fact that the buildings were not much changed since the book was written made it feel as if we had stepped back in time to the world of the story.

(Excursion Participant)[4]

Figure 11.1 Plotted Excursion for *The Owl Service*.

Walking through plot-points from the novels, alongside an assemblage of corresponding locations, appeared to add new dimensions to the literary geographies of plotted readings. Participants could develop new insights into the cultural and physical geographies of these borderlands, sensing the sights, sounds, and emotions evoked through this direct encounter, whilst recalling characters from the novels and their own experiences of reading these plot-points. Through this process, participants sensed a new connection to and identification with the locations being moved through:

> Seeing locations and experiencing their atmosphere was great. Bethesda would have been 'just' another place if I hadn't read *Pigeon*, now it gives an extra layer to visiting Bethesda, but also visiting Bethesda adds an extra layer to reading the book.
>
> (Excursion Participant)

> But below the grey domain of this hill, the patchwork fields stretch their expanse of emerald down, sloping to a silver sea of torn paper waves. And above, above the hill, there are the crouching mountains, with their lakes, like broken mirrors wedged between valleys, and along the tops of the plaited ridges there's that trembling, pencil-line horizon. It's worth a second look. Just briefly.
>
> (Conran, 2016:14)

Figure 11.2 Conran having a "second look" at the plot of *Pigeon*.

Figure 11.3 Plotted Excursion for *The Revenant.*

> The places themselves now connect to how I read the books. I don't think I can go to a section of *Revenant* [for example] now without thinking of the place I visited, even if it is just a mention like the Clock Tower. The rainbow houses and even the little side streets around the manor are now different even though the words haven't changed.
>
> (Excursion Participant)

> just as they're going to, he points across to the clock on the church steeple and tells them he's putting a curse on it, that it'll never run on time again. And it didn't either, she tells us. It's been five minutes late ever since.
>
> (Hughes, 2008:28)

> The sharing of passages from the book brought it all to life.
>
> (Excursion Participant)

We have seen how, in Chapter 10, the experience of engaging with the broader context of the selected novels on the *Literary Atlas* website made respondents "think more about the interactions between text and world" (Reading Group Participant), in effect, making stories more "real" to them. This sense of dissolution of the modern categories of "real" and "imagined" was felt even more intensely through the plotted excursions. The locations as crafted through the words of the author, the experience of a plotted reading, and the corresponding geographies walked through in person gave the sensation that these components were not simply assembled together but had "converged" in practice. The notion of "convergence" has been previously developed in work on human engagements with post-nature (Anderson, 2009) and surfing spaces (Anderson, 2012, 2023). This concept refers to a co-ingredience felt by a human actor when they feel a sense of community and co-constitution within a broader assembly of entities/processes (perhaps in terms of non-human species, landscapes, and weather (see Anderson, 2009), or wetsuit, surfboard, water, wind, continental shelf, energy transfer, sun, in surf-riding (see Anderson, 2023)). This sense of convergence resonates strongly with the feelings expressed by participants on plotted excursions. Reflecting the multidimensional nature of the notion of plot (where the drama and geography of the crafted word mutually inform and define each other), and as plots in turn become more complex when merging with a range of corresponding geographies called forth by a reader, excursions developed this assembly further as a reader is brought into embodied engagement with their plots. As this aggregated borderland takes shape, the reader's plots are sensed as "being brought to life" (as a participant noted earlier). In this way, literary geographies become a converged borderland of plot, with participants experiencing the sensation that there is now no ontological or otherwise categorical difference between the material location of the plotted excursion (e.g. Bethesda), the location crafted by the author (e.g. Conran), and their own plotted reading (the individual's corresponding geographies). This coming together is not experienced as a modern locking together or connection of separate entities within a passive context, rather it is felt as a convergence of once separate experiences and processes into a new state of mutual interference and interdependence. Through this experience, what we understand as plot is now sensed as an emergent, multidimensional, co-equal borderland.

(De)composing plots

Despite being a convergent experience for many, it was not inevitable that plotted excursions affected all readers in this way. As stated earlier, there was always the potential within plotted excursions for these experiences to rupture the borderlands of some taking part. Indeed, for a small number of *Literary Atlas* participants, the corresponding geographies experienced through plotted excursions did not always strongly resonate with their own associations. For some, due perhaps to the passage of time between authorial composition and visit, the particular type of walking engagement, or even the weather on the date of the excursions, a resonant version of the chosen place did not emerge from its infinite variety of possibilities (see Solnit, 2010, but also perhaps Shakespeare's Antony and Cleopatra, Act II Scene II). For some, the encountered places therefore failed to correspond with how individual readers plotted them in their reading, as the following comments indicate:

> Having gone to the place made the story more my own, but by changing the experience to include something beyond my imagination revealed the novel to be more flawed in focus.
>
> (Excursion Participant)

> For me, St Seriol's Well was smaller and less impressive than I imagined, Penman Beach was brighter and not as oppressive as I'd imagined.
>
> (Excursion Participant)

It was possible therefore for some corresponding geographies to simply fail to align with some individuals' plotted reading of them, and in these circumstances, although the plotted world of the reader may be retained, the complex convergence with further geographies is forgone. Despite this, because plotted excursions were a collective practice, walking through plot-points nevertheless appeared to augment and consolidate the public sphere characteristics of reading groups (see Chapter 9). For example, plotted excursion participants remained open to sharing their own individual perceptions (even when they failed to resonate with words on paper, or other participants' views). More generally, participants commented on the simple pleasure of creating a space with others, often strangers, who had a shared experience of and commitment to the plotted reading of a particular novel. This communion of interest appeared to encourage an open-mindedness to difference and new interpretations of the plot-points in the borderlands – an embracing of tropophilia if you will (see Chapter 10) – as the following participants outline:

> I loved seeing specific locations involved in the novel, but also loved meeting with other people who have read the book and could voice opinions about the book and places visited.
>
> (Excursion Participant)

> I loved experiencing the chapel, the walk to the quarry, and having time to talk about the novel over lunch. I also loved meeting new people who had also read the book and shared an enthusiasm for the trip.
>
> (Excursion Participant)

> Sharing ideas and reactions in the group enriched my experience of reading the book.
>
> (Excursion Participant)

> I thoroughly enjoyed seeing the locations and readings in the book. It was fantastic to hear different opinions people have, and look for the locations mentioned . . . .we had lots to think about and great discussions.
>
> (Excursion Participant)

> I appreciated the opportunity to meet other people and discuss views, to consider other ways of interpreting the book. Listening to other people's experiences and hearing of other things going on in this area – it was seeing somewhere I've visited before, but not seen like this. Wonderful!
>
> (Excursion Participant)

Through the plotted excursions, participants enjoyed the process of bringing their own borderland into conversation, collaboration, and sometimes convergence, with others; due to their willingness to encounter and engage with disorientation of their person–place relations (see, for example, Chapter 8), they were open to the possibility of both resonance and rupture of their plotted readings. To illustrate this process of (dis)orientation through dialogue between participants, their plotted readings, and their experience of particular corresponding geographies, two extended transcriptions (firstly from *The Owl Service* excursion to Dinas Mawddwy, then, secondly, the *Shifts* excursion to Tredegar) will be shared. The following first extended excerpt goes some way to illustrate how a collage of collaboration occurs between the numerous dimensions of the now converged plot: the participants, their plotted readings of the novel, their own corresponding geographies, and the place of excursion itself.

F3: My first reaction is, it's not at all like I imagined. Not at all.
M2: Okay. so what did you imagine?
F3: Somehow, I thought we were higher up the hills, you know, with the house being sort of set into the hill. Is that how he describes it?
F4: Yes, that's right.
F6: I think it's more open than I expected it to be. I didn't feel quite as closed in as I was expecting.
F3: I think if you were down there by the river, you'd be very closed in, yes.
F5: I think it would, yes.
F3: It is a narrow valley, isn't it surrounded by very high hills.
F5: Well, I think that's why [Alison? name character] points out how easy would be to make this a reservoir.
M3: Yes, she does. You can just . . . a dam down at the end.

Figure 11.4 Plotted Excursion for *The Owl Service*.

M3: Yes, that's her way of thinking about it. And then you suddenly become really aware of depth then, don't you.

F3: And it is a reservoir for all the emotions and experiences too.

F2: I find these hills, they make you feel protected.

M3: Yes, its odd isn't it. It's kind of comforting almost.

F2: That mountain energy. Very protected, very safe.

F3: In a hot summer this place might well feel like a cauldron.

M3: Well, yes. exactly.

F3: It is, isn't it.

F6: So other people found it really, really claustrophobic.

M3: This is clearly what Garner picked up here.

F1: A line I remember was 'the mountains hung over him, ready to fill the valley'. You get that feeling even now, and on a hot day especially, if you can imagine it. But even now it's almost like this is a reservoir or whatever, something's ready to fill the valley. Maybe it's the mountains or the trapped violence and emotions and mythology.

M2: Say that line again that you just mentioned.

F1: That's the line that leapt out for me: 'The mountains hung over him, ready to fill the valley'.

M2: We're vulnerable. Fragile.

F1: I know. So it's all there, isn't it. But this is amazing, isn't it [gesturing to landscape]. You really do see it's mountain after mountain and nowhere to go.

Figure 11.5 Plotted Excursion for *The Owl Service*.

F1: Can you imagine feeling that you were totally being driven back because you can't leave. And it's his future.

M4: It's slightly eerie and macabre.

F1: It's a bit like that when Alison says . . . she's walking up what he [Garner] calls the peat road, and then when she can't see the house anymore, suddenly it begins to feel inimical. And I've had that feeling sometimes, not here, but in the Yorkshire Moors, North York Moors sometimes. You look out and it disappears as if you're so insignificant. And the whole of nature. And so you can get . . . if you were in desperation, pacing the mud, you just . . . you know, you've got . . .

F1: Nature's not always benign. It's quite often not benign.

M4: Yes.

F7: Also, at other times, you can feel the power of these mountains.

F1: Oh exactly. Other times it's really positive.

F7: And you can tune into them, and it can really take on that power and feel the other way.

F1: I remember thinking something like that. And I suddenly thought I'm panicking. And suddenly I thought, that's the route of the word: Pan, the God Pan.

F7: The God Pan, yes.

F1: Pan-ic. And I thought, yes, that's it. I really felt a kind of panic. It was totally irrational. And as soon as I started rationalising it and thinking about it, it was all right. But it was just that irrational moment of just feeling you were just a speck in a non-benign, a malign kind of atmosphere.

F4: And you carry your history with you everywhere, don't you. It's a very different sense of living . . . .But I just cannot imagine living in an area like this with that degree of isolation. And I was thinking, you know, I expected that when I came here I was going to feel that the environment was slightly menacing and a bit mysterious. And actually, I didn't feel that at all when we went down into the field. And I just thought, it's beautiful. It was just stunningly beautiful. And I didn't feel at all menaced. But when we went up further and there were all those black hills on this side, I found that very menacing. But then, when we got to the top, and then it was just glorious, wasn't it. Looking out for miles and miles and miles and miles and miles; that sense of openness. So I think the thing that I was left with was that actually, to me, I would feel very dislocated if I was living here.[5]

In this extended excerpt, we can identify how, in the "talking whilst walking" and "polylogic" tradition (see earlier), an unstructured dialogue emerges where all actors participate in a conversational, geographical, and informational plot creation (see Anderson, 2004; Anderson et al., 2010). The experiences of participants, their plotted readings of the novel, the corresponding geographies they bring to this reading, and the material geographies of this location, prompt a series of conversational forays that bring together a collective understanding of the plot of the novel. Given the power of this experience, and informed by other respondents (earlier), it

Figure 11.6 Plotted Excursion for *Shifts*.

is likely that, for these participants, this plotted excursion to the borderlands of *The Owl Service* will be strongly tied to Dinas Mawddwy (and vice versa); the senses and ideas prompted through these experiences will be similarly co-constituted (so, for example, when one may feel panic, this sensation will be somehow interconnected with Garner, the North York Moors, the Mabinogia, Mawddwy, and perhaps even the *Literary Atlas*). It is this coming together of words, experience, reading, landscape, literature, and conversation with others which produces a rich, textured assembly of borderlands and an active convergence of plot.

A second shorter, but nevertheless extended, excerpt comes from the plotted excursion to Tredegar for Christopher Meredith's *Shifts*. Prompted by their engagements with an old, but semi-protected ironworks, the participants' conversation actively brings together the crafted words of the author, his characters, and their own "real-time" plotted excursion to identify the difference between this form of creative tourism and more packaged versions of literary tourism:

F7: Look at this, it's just been, kind of, 'made safe' in quotes and walked away from really.

F3: 'Let's put a handrail up' and that's it [laughs].

F7: And cages over the door, but you could climb through that. (I mean I would be tempted to climb through and see what's inside, because I'm a . . . I tend to be a little but adventurous, which is not sensible but it won't stop me doing it).

F2: I'm not sure what to make of this. Because when you go to museums they frame it for you don't they? They tell you why it's important and why you're here, and there's a . . . people go because they want to see this important piece of whatever because it's celebrated that way, but [here] because there's no

> [official explanation] . . . to look at. Look, as you said, it looks abandoned I don't know what I'm supposed to think about it.
> [Group make sounds of agreement]

"Keith kept sober by not drinking on his own round and sometimes leaving nearly full pints when they moved to a different pub. By the time they got to the bar of the hotel, Jack was drunk and abusive, though it was too crowded and noisy for anyone to notice.

Still Keith felt his face redden when Jack turned to history.

'It's not something you can escape from', Keith tried to say.

As they spilled out through the double doors with the crowd after stop tap, Jack lunged across the pavement and missed his footing on the kerb.

The last slab. It was a non-event. There was no blizzard, no panniered donkeys, no mysteries. The place lost money, that was all. The last coil of the old hot mill would be the first through the new tinning line and memorial tins would be pressed from it. Yet it seemed to Keith that the changes were very big. They were bigger than he had ever suspected and he felt that all he had even then was only a glimmering of the real scale. It was like discovering a huge dark planet that had always been there affecting tides and spasms in the earth and its effects permeated everything and so it was invisible. It was like the people who worked in the piggery who couldn't smell the pigs. Odourless, invisible history would blow them all apart and they would hurtle away from each other through space and never really understand what had shifted them" (Meredith, 1988:210–211).

F3: I can understand why . . . I think Meredith hit it in a nutshell saying just that. What do I think when I look at this? I suppose if you go into an art gallery and it's art that, you know, like minimalism which many people don't take to, you know, you think what the hell am I supposed to think of this? You know? . . . [Laughing]

F2: But I think Keith is trying to do something very important, I think he's trying to do something noble.

F4: Like, like us on this trip, you know, this, this has awoken an interest in me.

M: And this is kind of what Keith is doing in the book isn't it?

F4: Yes.

M: He's trying to . . . he feels as if there's more to life . . . and in history or something, in looking to the past, you might be able to define yourself in the present, that's what he's trying to do . . . It's a, kind of, interesting provocative question isn't it?

F4: It just, it also emphasises the sadness in the book as well doesn't it? The, you know, the deprivation and the deterioration of their current circumstances.

F4: That's emphasised isn't it? By looking here.

In this extract, we can witness the plotted excursion participants encountering a corresponding geography to a plot-point in Meredith's *Shifts*, and reflecting on how this location ties into ideas of history, institutionalized curation and exposition, and the power of personal engagement with one's own person–place relations. This process constitutes a bringing together – perhaps even convergence – of what the modern terms the "real" and "imagined": a written text, a personal plotted reading, and imposed corresponding geography (by the *Literary Atlas* team), and the experience of these multiple layers coming together through walking experience. Through this process, a contemporary conversation emerges between crafted words, collective histories, personal associations, and geographical places which dissolve the meaningfulness of terms such as fact and fiction, and creates a borderland in which understanding and definition can be collectively identified.

Conclusion: "let's have more of these literary feasts, please" (Excursion Participant)

This chapter has introduced and undertaken an evaluation of the plotted excursion into the literary borderland. Through extending a plotted reading from crafted words to first-hand engagement with corresponding geographies, an active, embodied, and "stimulating" (Excursion Participant) experience can be made available to those involved. A number of key insights can be identified through this evaluation. Firstly, through engagement with the multiple components that assemble together on plotted excursions, participants affirmed and validated their own read-realities of their selected novels. Although not all corresponding geographies resonated with all readers, the collective experience conjured convergence for many – with modern categories of the "real" and "imagined" dissolving through interaction, interference, and collaboration between participants, readings, words, and places. Secondly, the active and emergent nature of these plotted excursions was dialogic and generative; rather than being premised on the passive consumption of information packaged by others, participants were willing (and indeed enjoyed) the process of interpreting and realizing a borderland space themselves. This insight is exemplified by the final except from a plotted excursion to borderlands of Raymond Williams' *Border Country*:

F2: I think a lot of today, people don't let their imagination go on a journey. They want quick fixes. They want a film with a big box of popcorn and whatever. But to sit in beauty . . . when I mean beauty I don't mean make-up and all that rubbish; I mean it's at one with yourself, with your mind, and you read such a beautiful novel like *Border Country*, I think it's a luxury. It's a luxury to dip into that we should all dip into more often.
[Group make sounds of agreement]

F6: I loved the book the first time I read it. The second time I read it [for the *Literary Atlas* book group], I loved it even more . . . . I've never been to Pandy, and [today] really brought that book to life for me – coming here, seeing the little places and the names and so on, and that each cottage is isolated and so on. That really brought it home to me.

F5: It's almost like if it were a film, [we're told:] this is how it is. There wouldn't be that room, would there, to imagine?

In this chapter, therefore, we have seen how plotted excursions can be understood as a new form of literary tourism, providing a space to collectively conjure a literary geographical borderland, bringing words and corresponding geographies together in practice. As we have seen, through conversation and collaboration between participants and their places, borderlands are created that merge together modern categories and multiple locations. As this occurs, the plotted excursion becomes a literary geography which comes to constitute and co-define how novels and places are understood by its participants. Given the time that excursion participants were willing to invest in these excursions (in terms of travel time, as well as excursion participation), it is clear that there is a significant level of interest in such experiences. Building on their enjoyment of the plotted reading of novels, and the ways this mobilized recognition of the importance of their own geographical experiences (not simply to their own lives, but in the generation of a functioning plot for the novel), individuals were "hungry" (to extend the metaphor chosen by a participant earlier) to "feast" on the physical experience of geographies which had been considered to correspond to these plots. Through doing so with others in the reading group tradition, plotted excursions became stimulating and rewarding encounters, creating connections between the literary, readers, and physical places, as well the histories and contemporary communities that defined them, together converging into the development of a co-ingredient sense of (literary) place, which through experience, came to be part of their own geographical condition. Through the promise demonstrated by these plotted excursions, it is suggestive that they could be harnessed by and for others; perhaps becoming a mechanism through which further experiences (be they pedagogic or purely touristic) could be engendered to foster awareness and understanding of place, history, and literature (see also Chapter 13).

Notes

1 If, as Allington (2011) state, these individuals are indeed predominantly derived from the middle classes (see Chapter 9), then this cohort's availability of disposable income and leisure time may to some extent account for their "significance" in the commercial development of the phenomenon.

2 In this way, the plotted excursions operated in the "talking whilst walking" and "poly-logic" tradition (see Anderson, 2004; Pink, 2007, and more broadly, Sheller and Urry, 2006; Binnie et al., 2007; de Certeau, 1984; Anderson, Bevan, Adey, reference), for more see below.

3 Participants had to be over 18 to attend and in practice, reflecting the constituency of the majority of reading and writing groups involved, the average age was just over 40 years old. Each trip was voluntary and unpaid, but travel expenses could be claimed by participants if required. Each event involved both locals based in close proximity to the setting of the novel, as well as those travelling up to 150 miles to participate.

4 The following excerpt from *The Owl Service* depicts Garner's characters treading in the footsteps of the plotted excursion:

'Come here, you strange girl,' said Gwyn, 'and listen. We've had a good laugh, and we're on top of the mountains, and it's a sunny day, and there's nothing to be frightened of. But you must listen, because we've got to go back soon.'

'I can see why these valleys make good reservoirs,' said Alison. 'All you have to do is put a dam across the bottom end.'

'Not the most tactful remark,' said Gwyn. 'But you're dead right.'

'It's quite a thought, though,' said Alison. 'That thin bit of silver down there would fill the whole valley in time, and we'd be sitting here on the edge of a lake. Clive was wrong! This would be just the spot for fishing – better than his old pools, anyway.'

'Have you noticed how you can hear the river, even though it's so far off?' said Gwyn. 'And the motorbike going up the pass? Sound rises. Listen to that river. It's what lasts. Wherever you go you can think of that noise, and you know what you hear in your head is in the valley at the same moment. It never stops. It never has stopped since it began. It was the last sound Lleu Llaw Gyffes heard before he was killed. Gronw heard it, in his turn. We hear it now'.

'Gwyn-'

'Shh. Don't be frightened. Listen'.

(Garner, 2002 edition:139–140)

5 This conversation has uncanny echoes of R S Thomas' poem "Reservoirs" (1968, see All Poetry, no date), particularly with respect to the broader politics circling within Garner's rewriting (or perhaps for some scavenging) of ancient Welsh myths:

There are places in Wales I don't go:
Reservoirs that are the subconscious
Of a people . . .
. . . the English
Scavenging among the remains
Of our culture . . .

12 Engaging publics

Evaluating the *Literary Atlas* exhibitions

As the previous chapters have outlined, the Literary Atlas project was committed to the evaluation of our plotted reading approach, assessing how it may be generative for users within and beyond the academy. As Chapter 7 has discussed, the *Literary Atlas* team also commissioned a range of artists to visually interpret their plotted readings of the novels and to innovatively represent these borderlands. Although these artworks would be shared with the public via the *Literary Atlas* website (www.literaryatlwas.wales/artworks), the project also wished to directly engage with stakeholders via the physical exhibition of these artworks. These exhibitions would provide the basis for a series of public lectures and workshops, directly seeking to engage stakeholders who may not normally be strongly associated with what some have described as the "reading class" (see Chapter 9). The aim of this engagement was to gain insight from as broad a constituency as possible about the *Literary Atlas* project, and encourage individuals to participate in the reading, writing, and mapping of their own plots in the exhibition- and workshop-spaces. To that end, a series of public lectures, exhibitions, and workshops were put into operation, in 6 gallery spaces across Wales. This chapter presents the details of this engagement, centring around audience responses to the *Literary Atlas* materials in aggregate, as well as the work and feedback produced from the accompanying public workshops.

Curating literary geographies

Mobilized by the relational sea change across the social sciences, arts, and humanities, and the currents of culture, spatial, cartography, and creativity that synergistically combine with them, the *Literary Atlas* project was keen to pursue the extension of literary geographies away from artefactual engagement and towards emergent activity. Aligned with the arguments made in Chapter 3, it was clear that the subject areas of curation and museum studies were also not immune to these impulses, becoming sensitive to "the work that exhibitions themselves do, on and through audiences" (Ferguson, 1996:178). Kirshenblatt-Gimblett describes the influence of these currents as transitioning curation away from an "informing" paradigm and towards a model of "performing museology" (2000:10); through so doing, emphasis has been reoriented "from the museum as a collection of objects

DOI: 10.4324/9781003321248-12

[and] to the museum as a site of social and corporeal practices" (Leahy, 2012:3). As such, museums, galleries, and exhibition locations are evolving into "spaces . . . where affectivity is enacted and social relations may be transformed" (Thobo-Carlsen, 2016).

However, this paradigm shift is occurring within considerable structuring parameters. Many galleries have strict operating objectives (often set by their funding bodies), tight budgets, lengthy (12–24 months and beyond) timelines for space availability, as well as limited resources for curatorial innovation (see, for example, Gray, 2010; Folke Henningson et al., 2020; Brown, 2022). In this context, it was unclear what role the *Literary Atlas* team could take in this changing model of curatorial operation[1]: were we to be content providers or collaborative partners?

Compounding such uncertainty, we also anticipated a range of practical issues in engaging with exhibition spaces. The tyranny of wall space and size, the problems of quick turnarounds, the necessity of labour and expertise in storing, transporting, insuring, setting up, and arranging artworks, the need for and nature of (bilingual) exposition, the provision of audiovisual technology (to gallery specification), the need for catalogue creation and publication, exhibition advertising, and workshop recruitment and tutoring, all presented key problems to overcome. Compounding these issues, we were acutely sensitive to the methodological problems facing us when making a commitment to evaluate possible public engagement with these spaces. How might we appropriately capture feedback from potential users? We were intrigued by the possibilities of measuring participant responses using GSR and GPS technologies (see, for example, Tröndle et al., 2014; Hoare, 2019), yet we were also aware of the contemporary (and often disabling) limits to these potential advances. Thus, despite leading scholars in the subject of geohumanities (see Hawkins, 2011, 2012, 2021), suggesting that for some academics art engagements "had perhaps become . . . inherently valued as research" (Hawkins, 2021:1), we were nevertheless aware of the multiplicity of problems faced, and the lack of institutional infrastructure available to tackle them.[2] Despite this, the commitment to this work remained a beguiling extension of not only *Literary Atlas*' mapping innovations but also our wish to share the outcomes of our plotted approach with new publics. In this spirit, we approached a range of gallery spaces to consider the possible alignment of interests, objectives, resources, and timelines.

Homing the *Literary Atlas*

The *Literary Atlas* team approached a number of galleries with a view to establish sites for lectures, exhibitions of mapping materials (including commissioned cartographies), and public workshops. A range of gallery spaces, broadly covering the geographical spread of Wales, had objectives that aligned with the aims of *Literary Atlas*, and a timeline of events encompassing a launch, conference engagement, author events, plotted excursions, and a touring exhibition was put into operation at the following sites: The National Museum Wales (Cardiff), Oriel Davies Gallery (Newtown), Galeri (Caernarfon), the National Library of Wales (Aberystwyth), Pontio (Bangor), Swansea Museum (Swansea), and the Senedd[3] (Cardiff). The majority of these spaces could house the 12 cartographies commissioned by the *Literary Atlas* (see Chapter 7), along with the array of other materials, maps, and information required to contextualize these artworks.[4] None had the technological capacity to track visitor journeys around their spaces. When working with these venues, the *Literary Atlas* in essence became a content provider, filling the space of the gallery with materials which were felt to interest their audiences. How each space was set up was often dictated simply by the size of artworks and the available wall space, the capacity of temporary walls to be established to create audio-visual rooms, the availability of display cases, wall-hanging facilities, projectors and public-facing computers, and the desire for spaces of contemplation where the visitors could sit, engage with, and reflect upon the project in general. From this basis, many galleries sought to generate bespoke activities which resonated with the presence of *Literary Atlas* (for example, providing materials to encourage local mapping and writing exercises which synergized with *Literary Atlas* workshops). These activities coupled with more formal feedback opportunities, with a range of guest books, response cards, and surveys (both online and on paper), to capture user comments about the materials presented.

Public workshops were also established to coincide with the *Literary Atlas* exhibitions at these venues. Workshops were especially designed to reach those who may already be aware of, and interested in, gallery spaces and their contents, but also those for audiences who may lie beyond what is often called the "reading class"

Figure 12.1 *Literary Atlas* Exhibition at National Library of Wales, Aberystwyth (1).

Figure 12.2 Literary Atlas Exhibition at Oriel Gallery, Newtown (1).

Figure 12.3 Literary Atlas Exhibition at Oriel Gallery, Newtown (2).

Figure 12.4 Literary Atlas Exhibition at National Library of Wales, Aberystwyth (2).

Figure 12.5 Literary Atlas Exhibition at Oriel Gallery, Newtown (3).

Figure 12.6 *Literary Atlas* Exhibition at Senedd, Cardiff (1).

Figure 12.7 *Literary Atlas* Exhibition at Senedd, Cardiff (2).

(see Chapter 9). The latter was achieved through workshops accessed through specific invitation, principally issued to local English- and Welsh-medium schools, at both primary and secondary levels, and including all learning needs. *Literary Atlas* recruited a bilingual expert in artistic workshops (Sian Northey, also mentioned by Iwan Bala in Chapter 7) as a tutor to set up and run these activities. Along with gallery-run events, 40 activities took place in total during the touring exhibition.

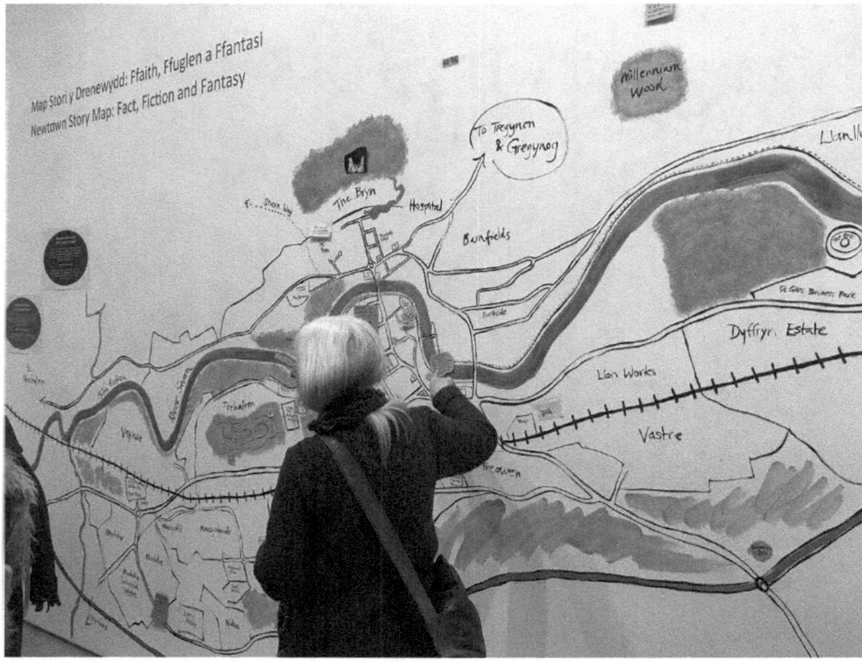

Figure 12.8 Literary Atlas Workshops at Oriel Gallery, Newtown.

Over 500 attendees participated in these workshops and ancillary events, and over 20,000 people attended the *Literary Atlas* exhibition in the 6 venues across Wales.

Exhibition reviews

Unlike traditional scholarly outputs which are written for, and (given pay-wall restrictions) often limited to specific disciplinary publics, gallery exhibitions garner reviews from beyond the academy. An illustrative cross section of media reviews on the *Literary Atlas* exhibition is given as follows, first from *Buzz Magazine*, and then from *Planet Magazine*:

> Each artist featured in this exhibition has a completely different style, the result of which is a truly contrasting collection of the cartographic imaginary of Wales. Measuring the Distance by Rhian Thomas is one of the pieces created for this exhibition. Responding to *Border Country* by Raymond Williams, Thomas takes the semi-autobiographical story of the Price's family's life in a small Welsh village and conveys the power of human attachment to a given place. Thomas' background is in architecture, rather than fine art, resulting in a unique style that bridges the gap between design and creation. Through the tearing and layering of historic maps, she creates a work

> that is carefully curated yet wonderfully abstract. The multidimensional craft includes a train line, symbolising the novel's main character Matthew, whose struggle to distinguish between the current awareness of his environment and the romanticised memories he holds, is a key theme of the book.
>
> (Urquhart, 2019)

> The twelve artworks are a meditation on th[e] spatial experience of Wales, exploring place and its multiple, mutable meanings. Yet, as '*Literary Atlas*' suggests, what is equally significant is the interplay between fiction and visual art. The project pairs twelve significant Welsh novels in English with a visual artist, asking them to respond creatively to the story and its setting.
>
> The result is an astounding, intricate, and multi-layered series of works that pay tribute to the complexity of both the novelists' and the artists' vision . . . .this outstanding exhibition demonstrates the parallels between creative processes, the siblinghood of literature and art, and . . . the possibilities of future interdisciplinary exploration.
>
> (Gramich, 2019)

Alongside these media reviews, the public themselves offered their own evaluation of the *Literary Atlas* materials shared through the exhibition spaces. As noted earlier, given the influence of the relational sea change and associated currents, exhibitions are seeking to become spaces in which "affectivity is enacted and social relations may be transformed" Thobo-Carlsen (2016). From over 200 written comments, it was clear that the visitors of the *Literary Atlas* exhibition gained an affective – and positive – experience,[5] and chose to articulate it in the following terms:

> Thank you for this wonderful exhibition, it brings together the maps, mapmakers, history, travels, connections.
>
> (Exhibition Attendee)

> Art exhibition unique!
>
> (Exhibition Attendee)

> "Fantastic collection"/"Wonderful collection"/"Great works!"/"Excellent"
>
> (Exhibition Attendees)

> Great ideas – deep mapping with a twist!
>
> (Exhibition Attendee)

> Good idea with plenty of supporting info. Good to see copies of the books!
>
> (Exhibition Attendee)

> Brilliant concept and execution – loved the video!
>
> (Exhibition Attendee)

> *Literary Atlas* is a fascinating exhibition. Congratulations to all involved.
> (Exhibition Attendee)

> Original and fascinating exhibition.
> (Exhibition Attendee)

> Loved the book recommendations and what a great idea to produce artistic interpretations.
> (Exhibition Attendee)

> Bigger and better than had imagined.
> (Exhibition Attendee)

> Evocative, mysterious layering and discovering. An impressive approach. Don't you *love* mixed media??
> (Exhibition Attendee)

> Llongyfarchiadau!
> (Exhibition Attendee)

Amongst the positive reviews, many visitors felt it important to pick out particular artworks that affected them:

> Superb evocation of the experience of engaging with a novel – in particular 'Border Country'.
> (Exhibition Attendee)

> Joni Smith – very clever!
> (Exhibition Attendee)

> I really like the Amy Sterly 'Missing Piece' it really is unique.
> (Exhibition Attendee)

> The work of John Abell really caught my eye, I like the way he brought everything together. Such an amazing colourful piece of art.
> (Exhibition Attendee)

> Love this project from start to finish – you are an amazing bunch – The Abell piece is incredibly stunning!
> (Exhibition Attendee)

> 'Concrete Ribbon Road' was beautifully crafted and touched me.
> (Exhibition Attendee)

> Nice Murder wall
>
> (Exhibition Attendee)

> Love it! Joni Smith caught my eye, very good.
>
> (Exhibition Attendee)

> I love George Sfougera's' [sic] work as you can see it from multiple perspectives. At a distance you see a face, but when you look closer you can see the story being portrayed. How he stained the paper so perfectly is also just incredible. The best artwork I have seen!
>
> (Exhibition Attendee)

> Loved the 'Sitelines' animation. Very relaxing and great colour scheme!
>
> (Exhibition Attendee)

> Beautiful photography, music and pure voice in Sitelines 'short'. Excellent and informative exhibition.
>
> (Exhibition Attendee)

The specific articulation of the "affect" sensed by visitors was notable. Many visitors expressed their "love" for the artworks (and specific artworks in particular), choosing to articulate their feelings in the face of these works through expressions of wonder or enchantment. Such responses suggest that the *Literary Atlas* materials appear to speak to these audiences in a specific way. Firstly, and as we have seen in the book to date, the plotted approach emerges from recognition of the integral relations between people and place, and furthermore, the umbilical connection between locations and stories. Starting from this premise appears to help gallery visitors relate to the *Literary Atlas* materials – these materials are referring to the human (spatial) condition that each visitor themselves has in their own ways; in this way, the *Literary Atlas* is not built upon (just) an abstract concept or an intellectualized theory, but rather on a human experience that we all share. This shared experience establishes further resonance if and when the geographies referred to in the *Atlas* materials correspond in some way to the visitors' own (at the scale of Wales, or at a more local level with the plot-points of particular novels). Secondly, users appeared to appreciate the *Literary Atlas* due to the combination (and lack of hierarchy) of media used in the exhibitions, as one respondent put it: "We are loving this immersion in landscape, culture, literature and art" (Exhibition Attendee). The coming together of the different cultural traces (for example, art and mapping, place and literature, but also the different genres of literature, and different media chosen for cartographic innovation (e.g. still or moving image)) seemed to have a powerful affect, offering different lines of flight into the *Literary Atlas* materials, but although through combination (and perhaps even convergence), offering the potential to be "thought-provoking", "inspirational", and "touching" in its affect (Exhibition Attendees). It is important to note this range of affective responses, in line with plotted reading, which imposes no modern categorization between processes of rational cognition and more open experiences of wonderment, combined with both intellectual engagement and also emotional reaction. As the following comments suggest, "this exhibition is where art and soul meet" (Exhibition Attendee) and "deep thoughts reflect great work" (Exhibition Attendee). Uniting intellect and senses in one set of plotted materials thus appear to provide something that was "fascinating", "stimulating", and for some even "mesmerizing" (Exhibition Attendees). It was as if the (literal) exhibition of passion for and in this subject area gave permission to the visitors to articulate their own passion for it too; but also perhaps that acknowledging the existence of borderland space emergent from the power of reading struck a powerful chord amongst visitors, identifying but also celebrating a space which for many forms a vital dimension of their lives.

However, as we have seen earlier, in recent years, exhibitions are not simply seeking to become spaces in which "affectivity is enacted" but also arenas in which "social relations may be transformed" (Thobo-Carlsen, 2016). One of the key ways in which the *Literary Atlas* had a generative effect (rather than an emotional affect) was through encouraging visitors to renew their acquaintance with books. As the following visitors comment:

> Very interesting and different artwork. It makes me want to read the books the pieces of are based on!
>
> (Exhibition Attendee)

> I had only read two of the novels but I intend to make a reading list of the others.
>
> (Exhibition Attendee)

> An interesting mixture of responses. Will now read the books.
>
> (Exhibition Attendee)

> The creative links between different art forms (here the novel and visual art) are endlessly fascinating. This exhibition reminded me of books to re-read and introduced me to writers and artists new to me. Very valuable.
>
> (Exhibition Attendee)

> Very interesting exhibition. It also made me want to buy a couple of the books referred to!
>
> (Exhibition Attendee)

It was clear that through visiting the exhibition and engaging (primarily) with the commissioned artworks, individuals became inspired to engage directly with literature. For some, this would be a return to literature, and for others an encouragement to begin reading, yet it appeared that the plotted approach, and the materials emergent from it, mobilized many to begin a new literary journey, a venture which would have entertaining, educational, as well as potential economic impacts (through, for example, books purchased). In this way, the plotted reading materials went some way to not only speak to the person–place relations that all visitors experience but also consolidate them through encouraging the act of reading, enabling individuals to find out more about their own and others' co-constituted geographies.

Participants in the public workshops also articulated a sense of mobilization from the plotted approach to the events, expressing a new interest in both reading and mapping and writing:

> This was a superb workshop. [The tutor] shared her enthusiasm for the *Literary Atlas* project and encouraged each participant with insightful and imaginative prompts.
>
> (Workshop Participant)

> Thank you for encouraging my creativity.
>
> (Workshop Participant)

> I enjoyed today so much. I learnt so much and it has inspired me to write.
>
> (Workshop Participant)

> I enjoyed the gathering of writers. I was stretched, unexpectedly, sometime scarily, through little encouragements and ideas. Thank you!
>
> (Workshop Participant)

> 'To educate' means 'to encourage forth from within', and this workshop encouraged our inner writer. Thank you!
>
> (Workshop Participant)

The *Literary Atlas* workshops appeared to provide the opportunity for their participants to creatively express their own literary and geographical connections through producing maps of all sorts. These workshops gave license to personal expression, offered an affirmation of skill, and provided a safe space for these elements of identity to be ignited (through the exhibition), and then explored and articulated (in the workshops), and were thoroughly appreciated by those involved.

Indeed, simply attending these spaces helped raise the profile of the gallery spaces for those who do not usually attend; the direct invitation to the workshops seems to generate an interest in the galleries more generally, as one gallery manager suggests:

> One interesting thing was that the teacher and assistants said that they never come to the [gallery], or if they do only to the café, but they enjoyed looking at the exhibition and were surprised at the range of activities it offers.
>
> (Gallery Manager)

Bringing new participants to the galleries, through exhibition attendance and workshop participation, generates new means through which individuals can become part of the community, learning about local and national culture, and perhaps even contributing to it in a small way – becoming part of the ongoing composition of their place, as one curator put it:

> It's been a really great show and we've enjoyed the dialogue it has brought about, thank you for bringing it to us.
>
> (Gallery Manager)

One example of a small but significant contribution – for the individuals involved, but also perhaps the community as a whole, was noted by the workshop tutor. The *Literary Atlas* materials inspired participants, in this case, young people with a range of learning specificities, to write and reconsider their world through working with words:

> Their written skills [may have been] varied, with the weakest only producing a few words with help, but this was more than made up for by their genuine interest in the artwork and in the books . . . Here's a class poem we wrote about their school:
>
> *A nice big sofa,*
> *A hard piano and a normal guitar,*
> *McDonald's, my favourite place,*
> *A healthy meal and also sweets,*
> *Sunny, snow, music weather,*
> *A game of football,*
> *A train track that's easy to travel on*
> *and a road that can take you anywhere.*
>
> (Workshop Tutor)

The creation of such lines in the workshop space can be seen to be a powerful statement; the "road" from their school, "which can take you [and them] anywhere", may refer to a material apron of asphalt, or a narrative "path through the woods" (after Pullman, 2017:87) of a story, but these lines demonstrate how, for the time and space of their involvement with the *Literary Atlas* exhibition and workshop, these participants had their creativity encouraged, their horizons widened, and their identities as articulate and intelligent members of a community affirmed. It was clear that the *Literary Atlas* touring exhibition and workshops accessed a significant number of individuals who may have never encountered the project if it were not for these public events. It moved many, and mobilized others, into becoming more interested in the English language novels of their locality and country, and empowered them to express their identity (tied to these places and histories) in a number of creative media. As the final chapter will go on to summarize, such consequences are significant: as one primary school child commented on leaving a workshop: "it taught me that ideas are powerful."

Notes

1 Fascinated by possibilities of engaging with broader publics in new ways, the *Literary Atlas* team had begun to tentatively explore how the spaces of a gallery might be used to emphasize the notion of plot, and help to tell a story spatially within these buildings. For example, we imagined that gallery spaces could be woven into a plotline journey around Wales, with floor, wall, and corridor spaces suggestive of different areas of Wales, with plotlines intersecting, and visitors moving through a labyrinth of plot-points and their associated materials. Yet we had no real understanding about whether such imaginings may be possible given the parameters faced by many exhibition spaces in practice.

2 It is important to note that given the commissioning aspect of the *Literary Atlas* project (see Chapter 7) was not part of the original bid for AHRC monies, all resources (in all senses) for this component, its display, and evaluation had to be obtained from outside this financial and temporal envelope.

3 The exhibition, talks, and tours in the Senedd were sponsored by the sitting Chair of the Culture, Welsh Language and Communications Committee, Bethan Sayed AM, and occurred in both the Senedd and Pierhead Buildings.

4 Given space restrictions at Pontio, Bangor, all artworks were exhibited digitally over large-scale projection screens.

5 Some venues conducted their own surveys of attendees. One such, Oriel Newtown, received 6269 visits across 32 days, which they considered to be a "good average" of 196 per day.

92% of those attending said the exhibition of the *Literary Atlas* materials "enhanced their experience", 84% agreed their visit was "enjoyable", and 81% agreed they "learned something new". The venue also offered the following categories in which visitors could articulate their responses to the exhibition: 23% suggested the exhibition "improved my knowledge or understanding"; 23% suggested they "felt a strong sense of personal connection and was moved emotionally"; and 31% said it had "stimulated my own creativity".

13 Conclusion

"Touch life often": the plotted approach to literary geographies

Introduction

> we live in a densely . . . storiated landscape; how do you find new ways of seeing those?
>
> (MacFarlane, in British Library, 2012b)

As we have seen in this book, the *Literary Atlas* project emerged as a response to the need to "find new ways of seeing . . . storied landscapes" (after MacFarlane, 2013). Drawing on work from Anderson (2014), *Literary Atlas'* "new way of seeing" was based on the plotted approach. Taken broadly, the plotted approach can be understood as one example amongst many across the social sciences, arts, and humanities which seeks to move academic enquiry beyond the "false oppositions" of the modern constitution, and principally with respect to literary geographies, the division between "text" and "world" (after Bennett and Royle, 2016:36). This false opposition has had clear consequences for the subject area, structuring scholarly work into the disembodied analysis of a "text" and the tracing of literary circulations between "the *lived* to the *imagined world*, and the *imagined* to the *lived world*" (Anderson, 2015:117, new emphasis). As it divided and opposed, the modern constitution effectively removed the potential for these categories of experience to be intimately connected or thoroughly co-ingredient in their nature.

Extending and developing the plotted approach became the basis from which *Literary Atlas* (re)conceived literary geographies as if they had "never been modern" (after Latour, 1993). In this spirit, the plotted approach moved away from analyzing texts or exploring circulations, but rather attended to how reading books can "produce our reality" (Bennett and Royle, 2016:36). Through doing so, it opened up a range of lively new outcomes and agendas for literary geographies.

To this end, the *Literary Atlas* project recorded an online "book of maps" (www.literaryatlas.wales) which located over 570 English-language novels set in Wales. Then, focusing on 12 novels in particular, it adopted a distinctive "plotted" approach to explore those novels and create a range of innovative cartographies based upon this exploration. As this book has documented, a range of maps were produced through this approach, emergent from the identification of plot-points, their relation

DOI: 10.4324/9781003321248-13

to broader geographical contexts, and the innovative expression of these borderlands of read-reality. To evaluate this range of contrasting and complementary maps, the project undertook a series of qualitative research activities involving reading groups, excursions, digital surveys, exhibitions, and public workshops.

> Plot is about cause and effect . . . If a plot[ted reading] happens, there is a result.
>
> (Goldberg, 1991:223)

As Anderson (2014) laid out, plot refers not simply to the dramatic development within a fiction, the arcs of its characters, and the process of conflict and resolution, but also to the geographical dimensions in which that story occurs. Anderson (2014) insists that these two dimensions of plot are not merely coincidental or passively joined, but rather are directly co-ingredient. Adopting a plotted reading of literature therefore explicitly focuses attention on the interdependence of drama and geography, recognizing how both are directly affected by each other.

As this book has demonstrated, this plotted approach offers a new way to see the storied landscapes of which we are a part. Complementing many iterations of mapping literature, part of this approach involves the transposition of the crafted words of an author onto a digitized representative cartography. By juxtaposing these components in this way, the connection between this story and this place is made explicit. Yet specifically, the *Literary Atlas* project framed these relations in a particular way. As stated, the plotted approach argued that this connection between story and place is not simply an interesting coincidence between the "imagined" and the "lived worlds" (see Anderson, earlier), but rather they emphasize how these components are actually co-ingredient in their nature; in terms of the story, this drama could not have happened in the same way anywhere else, and thus in some sense, the places authored, facilitated, or limited the possibilities of this drama. Reciprocally, these places are also to some extent now co-defined by the stories that emerged from them. Thus, from a plotted perspective we must consider that we do not simply "live in densely storied landscapes" (after MacFarlane, earlier), but rather we inhabit worlds that are co-composed by stories, we live in hybridized compound "storied-landscapes" that are better understood as "plots". In this version of literary geographies, we consider worlds where drama and geography are co-constituent and interdependent in an emergent ontology of becoming.

Thus, through making a plotted reading "happen" (after Goldberg, earlier, but also resonating with Hones, 2008), a process is begun that changes what is "at stake" (Law, 2015:127) for literary geographies. Like all projects which seek to challenge the modern constitution, at a fundamental level, this approach seeks to change "what the world is" (Law, 2015). Mobilized by the philosophical sea change across the social sciences, arts, and humanities (see Chapter 3), *Literary Atlas* responded to the pressing question of "how we are going to configure the world" (Dewsbury, in Hawkins, 2021:45) by identifying an alternative, amodern, relationality. As we have seen in this book, relational configurations of the world recontextualize the "noun chunks" (after Laurier and Philo, 1999) and concomitant "false oppositions" of the modern constitution, and replace them with an interdependent, constitutive ontology. In this framing, "things" are always redefined through their coming together with every-"thing" else. Changing these stakes results in a different type of literary geographies; it brings about a change of emphases towards the convergence of "text" and "reader", and in particular on the generative potential of the performance of reading – both as an individual practice (be it as an academic scholar or a lay public reader) and in terms of collective reading (for example, through the phenomenon of book clubs and plotted excursions).

Synergizing with a range of conceptual currents sweeping across the social sciences, arts, and humanities, this recognition of relationality emphasizes that reading discards the modern conceptualization of books as "stable ontic entit[ies]" (after Kitchin and Dodge, 2007:341, cited in Hornsey, 2012:675), and replaces it with an understanding of read-books as "fundamentally ontogenetic" processes (Hornsey, 2012). Following Thumala Olive, this recognition focuses attention on the "contingent and generative interaction between [what the modern terms] book/text, reader and context" (Thumala Olave, 2021:103). More specifically, we have seen how reading literature from a plotted perspective serves to make explicit the ways in which the drama of the story emerges from and conjoins with its geographical location. With this emphasis, *Literary Atlas* has demonstrated that a plotted approach offers a new way to approach, engage, and connect individuals with literature. With these insights, if we take seriously Larsen's reflection that

> it is sometimes said that the reader meets the author halfway to the page, but I would say it is more like 0.001% author, 99.999% reader's architecture of imagination.
>
> (Larsen, 2018:168)

then any form of reading offers the potential for a book to "nurture . . . intimate connections between [the] text and the [readers' own] life" (Kuzmičová and Cremin, 2021:37). Any form of active reading will transform literature from the author's own words into a story,

> packed with associations from our own personal histories. We [will] see both the thing itself and all that we associate with it.
>
> (Storr, 2019:41)

As a consequence, read-literature in general can be seen to be not simply a fixed product of the author, but co-produced by the reader and their own situated knowledge; we will "read with our pasts" (Storr, 2019:41) and insert our "preconceptions; values; memories; connections; and emotions [as] an active part of the story" (Storr, 2019:55).

However, influenced by the spatial current that has influenced a range of disciplines cognate with literary geographies, this book has argued that any "reader" is not simply a social being but also a spatial one (or indeed many). As such, the

Literary Atlas project has demonstrated that when one adopts a plotted reading, individuals do not simply read with their "pasts" (as Storr suggested earlier), but also with all the geographies and spatialized experiences that co-define their identities in both the past and present. Adopting a plotted reading approach thus activates not only our awareness of how drama is indelibly connected with the geographies noted by the crafted words of an author, but through so doing, opens up the possibilities for further geographical connections to be sutured through and to a reader's "architecture of imagination" (after Larsen, earlier). Thus, when Larsen goes on to state that the "trick" for any author is "knowing what details to include" to prompt a reader's engagement (Larsen, 2018:168), a plotted approach suggests that this engagement is taken to a new level when "plausible" (after Pryce, in Anderson, 2014:213) or "telling" (after Bradbury, 2008:108) geographical detail in the broadest sense is used to provoke this process. Thus, to paraphrase Bradbury's Dr Faber (in *Fahrenheit 451*), a plotted reading approach can help us to understand how "good writers touch life often" (Bradbury, 2008:108).

A plotted approach can therefore generate new ways to understand the practice of writing. As we have seen, exploring a writer's own person–place relations can help to illuminate the role that these play in their motivations for writing, their process, and the emergence of their plots. It can also help to direct attention to the importance they place on having "the geography right" (in the words of Trezza Azzopardi, Interview with *Literary Atlas*), as we have seen in Chapter 8, this may be in terms of material locations and directions, or the cultural locale and languages of their particular version of a place. Such focus echoes interest in the role of location for creative writing, but goes further to enhance and emphasize it as the primary mobilizing factor in dramatic development (see, for example, Yeh and O'Reilly, 2022). By adopting a plotted approach, we can therefore begin to ask new questions for literary geographies, exploring what type of dramas and geographies authors seek to conjure through their crafted words, how particular insights from the cultural and spatial current can illuminate the dynamics of these plots, and influence how we come to consider and engage with the borderlands they form around us.

A plotted approach can also generate new ways to understand the practice of reading. By asking individuals and collectives to approach books with a sensitivity to the role of place in drama, it opens up the potential for explicit connection to be made between the broader person–place relations of the characters and the readers themselves. As a consequence, plotted reading can be understood, at least figuratively, as a geographical exercise (see Chapter 3, and Hones, 2014). Plotted reading can be seen to be a practice of casting off the temporary moorings that anchor the reader's person–place relations in their everyday lives, and embarking on a journey into new sets of socio-spatial relations which they will encounter through the author's crafted words. In these cases, normal person–place relations are temporarily released or loosened, with new connections established through the practice of reading. Yet when this occurs, none of these relations is ever fully detached; due to the active role of the reader in contributing to the plot, this new assemblage of locations and drama become knotted together, extending a sense of identification and relation with this new read-reality. As we have seen, the geographical exercise of reading can be understood therefore as a process of (dis- and re-) orientation, changing and challenging a reader's person–place relations, introducing new components, and stretching and twisting existing perspectives. This process of "orientation" can be considered to be similar to Hall's use of the term "orienteering" (i.e. the physical navigation across unfamiliar terrain, see 2004:15). For Hall, it is important that orienteering is considered as a practice that always involves "more personal, more historical, more associative, more metaphorical, [and] perhaps even more spiritual" dimensions than is conventionally considered (2004). For Hall, orientating involves,

> crashing through the larger landscapes of memory and experience and knowledge, trying to get a fix on where we are in a multitude of landscapes that together compose the grander scheme of things . . . [it is a practice which] asks not just where am I, but where do I fit in?
>
> (2004:15)

With Hall's notion of orienteering in mind, plotted reading becomes a process of dis- and reorienting the reader through their borderlands. Read-literature becomes part of the key "cultural knowledge" (to use Hall's term, 2004), which allows the reader "to know what I know, which is often another way of knowing where I am" (Hall, 2004:15). Thus, in the same way that Turchi may suggest that every writer is also, in some way, a cartographer (after Turchi, 2004), we may also say the same of every plotted reader: every plotted reader becomes a geographer, an explorer, and indeed a map-maker, as they embark upon a process of dis- and reorientating their lived geographies with the plot of the author; and, through so doing, this process helps them to find their place in the world.

Thus plotted reading can, to use the term coined by Watson in relation to map orientation, produce a "special form of reading game" (2000:110, see also Pavlik, 2010). This reading game involves the plot-sensitive reader engaging with the crafted words of the author and producing new person–place relations which can change their world. This game may be played with a range of plots with which the reader may be more or less familiar. For familiar plots, where a reader's local geographies (for example) are included in the author's words, a sense of direct identification is experienced (it explicitly "touches their life" as Faber might have it, earlier). In these cases, it is not simply that the readers relate to the book (and thus an intimate connection may be generated in this way), but also that the crafted words of the author (sometimes echoing through ages from a writing process begun many generations before) relate to them. It is as if the author and their crafted words "see" the reader, and their own person–place relations (after MacFarlane earlier), and through this process an uncanny frisson is generated which alchemically conjoins the plot of the book with the plot of the reader's life. In this way, to paraphrase Storr, a plotted reading can "create a resonance that has the power to shake the [reader] as only art can" (2019:55).

When the nature of plots may be unfamiliar to the reader, with the individual or group encountering situations, circumstances, time periods, and characters from beyond their normal life, a plotted reading emphasizes how these encounters are experienced as a form of escape. In many cases, this extension of person–place relations is an enjoyable experience. The notion of tropophilia – or love of change (see Anderson and Erskine, 2014) – can be used to express how pleasurable and enchanting these re- and disorienting encounters are for many readers. Yet in some cases, the coming together of the crafted words of the author and the personal geographies brought into assemblage by the reader can be so disorientating that they "jar" the readers' person–place relations. This capacity may reflect the lack of "plausible" or "telling" detail used by the author in their crafted words – and therefore the reader may not consider it "good literature" as it fails to "touch the[ir] life" in a constructive way. However, it may also reflect a plausible but nevertheless uncomfortable expansion of their understanding (perhaps extending it to include aspects of crime, for example, that offer a darker vision of their world), and as such disrupt the inevitably precarious nature of person–place relations.

We have seen in this book that a plotted reading approach can help provide new ways of engaging with collective group discussion too. We have seen how book groups often involve an array of opinions with regard to the quality (or otherwise) of literature; as noted earlier, for different individuals with different person–place relations, the plot of literature has a diverse ability to "touch their life". What is common, however, is that by using the plotted approach readers become acutely aware of the infinite variety of outcomes that can be produced through the coming together of crafted words and corresponding geographies; as a consequence, readers replicate a relational acknowledgement that there is no "one" version of a place out there that we can come to know, and even when iterations are disorienting they nevertheless offer broadly valid contributions to this multiplicity. In this way, it can be seen that plotted reading to some extent reflects, but also actively (re)produces a world of read-relationality.

As such, the plotted approach is one example of an approach to literary geographies emanating from the working definition offered by this book. It offers a critical engagement with the geographical readings of fiction with a view to stimulate new understandings of the agency of this complex on (the cultural, spatial, cartographic, and creative nature of) the human condition. We have seen how, in the *Literary Atlas* project, that the plotted approach can identify ways in which literature can touch the life of the author and reader. By calling explicit attention to the geographical aspects of human life, these can be used to relate and connect to all aspects of the lived world, including authors, readers, and characters; it can help to generate new understanding of the cultural (b)orders effecting different time periods, circumstances, and situations, and help to generate affect and empathy for different others. It offers the potential for further questions to be raised in relation to what kinds of plot are produced by different authors, in different languages, in different eras, and through different forms of literature. This approach can be further extended by exploring how these plots are then taken and made by different audiences (by readers in different places, or with different approaches to learning and imagination), and through associated forms (e.g. how books are performed and projected through audio, through television or film production, and subsequently consumed and known through these different media), and how they may be marketed for tourism (be it creative, plotted, or otherwise). Through engaging in a plotted reading therefore, individuals and groups, and literary geographies itself, can be affirmed and reinvented in new, combined, geographies, cultures, and action.

New cartographies

As we have seen in the early chapters of this book, there has been a growing acceptance that all stories could be considered as some form of cartography, with the consequence of opening up new possibilities for the different purposes of a map, and the different roles of a map-maker (Chapters 1 and 3). Whilst traditional and new cartographies may both "give their makers the power to define the territory in their terms and write a singular version onto the landscape" (Owen, 2002:10), conventional cartographies often remain configured "as neutral carriers of information" (Owen, 2002.:011.) and their map-makers as objective charters of demonstrable truths. However, like many amodern exercises, a plotted reading challenges these conventions. The *Literary Atlas* project has demonstrated that it is possible for maps of all kinds to "*make* the relationships of things to one another real" (after Owen, 2002:11, emphasis added). For example, when converting one form of literary cartography (e.g. a book) into an abstracted bibliographic catalogue, and then into a conventional "distant" map, *Literary Atlas* used geoparsing to produce maps of key locations within bibliographic catalogues of Welsh writing in English (Chapter 5). Given the "slippery" nature of some locations, it was clear that some interpretation was required on behalf of the *Literary Atlas* team in order to tie the plot of the novel into corresponding geographies that were felt to be associated with the text. Similarly, with respect to our selected novels, our digitized "plot-points" became the basis for maps which outlined "regions" of literary geography in Wales. These regional maps served to produce new insights about the geographic distribution of the novel and the consequence of this for its drama. In its subsequent evaluations, it was seen how these distant maps could make the "relationships" of plot to place appear "real" (to use the words of Owen, earlier). These visualizations generated a new awareness of the diversity of cultural and more specifically literary areas of Wales and, through its modern rendering, these geographies were generally accepted as factual and objective.

Yet as the cultural, cartographic, and creative currents mobilizing the social sciences, arts, and humanities have demonstrated, all maps are authored arguments, rather than artefactual banks of objective fact. As such, map-making is a creative process, using a combination of visualizations and words to "make selections" for what the "story needs" (after Alan Garner, *Literary Atlas* interview with Author). In this way, cartography becomes displaced from its conventional scientific roots, and becomes entangled to some extent in artistic endeavour; the role of cartographer thus becomes entangled into a creative, philosophical, and explicitly political one. As author, as curator, and even as auteur, the cartographer becomes a role through which the academic subtly includes, excludes, steers, and charts a map in concord with the narrative that is being developed. In short, the cartographer "plots", and through this process a map (like a novel) is transformed from an artefactual repository into a recontextualized, considered, and situated lens through which reality is configured; through doing so, as Alan Garner suggests, "a map [becomes] a work of art" (after Garner, *Literary Atlas* interview with Author).

In this way, maps elide the dichotomy between cartography and literature and can begin to be considered as a cultural trace which also enjoys the capacity to "touch life often". Maps can offer new perspectives to their readers, stimulating, limiting (after Caquard and Bryne, 2009), and otherwise orienteering them (after Hall, earlier) in relation to their lived world. This recognition prompts further reflection on how academics should utilize these opportunities of creative cartography: what forms of knowing can (and should) be mobilized through map-makings? As the *Literary Atlas* project has demonstrated, these questions may not have one answer; it is possible for academics to deliver (facsimiles of) fixed truths, offer situated thoughts and reflections, embrace ambiguity, stimulate affect, or even mobilize action. It is possible for academics to produce a range of maps as a portfolio of "critically informed stories with the power to illustrate, illuminate, inspire, and mobilize readers to think and act critically and reflexively" (Denzin, 1998). As we have seen in this book, when users engaged with the cartographies of the *Literary Atlas*, it was clear that these not only stimulated a strong recognition that these stories were situated but also an interest in where this situating occurred; in other words, it sparked users' compulsion to embark upon a "spatial adventure" (after Casto Broto and Baker, cited in Bridge, 2018:1 in relation to disciplinary endeavour). These plotted maps prompted readers to question how the stories with which they were familiar had been connected to different locations by the *Literary Atlas* team, and by extension, how these connections interacted with their own experiences of these places. They were prompted to reflect upon their own knowledge of this place, this story, and the new conversations that were prompted as a consequence. They questioned whether, from their own experience, this location was a plausible choice for this story; they reflected on how this juxtaposition re-positioned what they knew about this place, this story, this plot, and how this (dis) oriented them. In short, they began an embodied and cognitive reflection of how these plots of literary geographies affected the plot of their own life.

The *Literary Atlas* cartographies thus demonstrated that the plotted approach to literary geographies had the capacity to stimulate an audience. This occurred with respect to Moretti's desire for individuals to "engage with a map and think" (1998:7), but more overwhelmingly, individuals engaged with these cartographies and *felt*. In the same way that Fletcher senses in relation to creative engagements with literature (2021), the appeal of the *Literary Atlas's* plots was more "visceral" than purely cognitive (Fletcher, 2021:55), touching life in ways that "we can feel the meaning even when we can't fully comprehend it" (Fletcher, 2021:58). In line with feminist currents within the social sciences (see Haraway, 1988; McDowell, 1999; Bondi, 1990, 1998, 2005; Pratt, 2004; Bondi et al., 2005b), these engagements remind us that, like academics, readers are no longer wholly modern, but humans with politics, emotion, rigour, and imagination and as such, respond differently when prompted to do so by the nature of various traces and the engagement opportunities they encourage. By adopting a plotted approach to literary geographies, it was as if users were compelled to draw upon their own person–place relations when engaging with our plotted atlas, and as such, it engaged them in affective, moving ways, as Goldberg's words (in relation to art of writing) put it, "the world" for *Literary Atlas* users "was now a little more alive", and these maps "made [them] a little more awake" (Goldberg, 1991:202/3).

New actors and new audiences

The plotted approach thus emphasizes how the sea change towards relationality and the associated confluence of cultural, spatial, cartographic, and creative currents offers the perfect storm conditions for literary geographies to appeal to many audiences, be they academic or public in nature. Counter to the growing consensus which suggests that literary geography (notably in the singular) is and should be an emergent discipline forged through the nexus of its two titular subject areas (namely human geography and literary studies), this book offers an alternative reading which acknowledges how scholars working in a multiplicity of modern disciplines are now interested in the subject (in the plural). Just as academics are seeking to work out from the "zombie categories" (after Beck and Beck-Gernsheim, 2002[1]) of the modern constitution, it is suggested that these scholars can also be seen to be working out from what may be seen as the "zombie disciplines" of academic organization. Through sometimes "turbulent and risky" (Beck, 1994:13) practices, which seek to "produce, stage and cobble together" (Beck and Beck-Gernsheim, 2002) new forms of interdisciplinarity, these scholars hope to "avoid strict typological categorisations" (following Hamlin and Speer, 2017:799) and not "risk de-politicizing" their analyses (Hamlin and Speer, 2017, see also Chapter 2).

Identifying, let alone inhabiting such a "difficult and inherently unstable space of betweenness" (Katz, 1994:67) is not an easy or safe option to take. Through doing so, projects such as the *Literary Atlas* risk alienating those practitioners who seek a more "modern" future for literary geographies. However, working out from the modern is a necessary move in order to explore the radical potential of literary geographies, both in a scholarly sense and also in terms of its public audience. The *Literary Atlas* has proven that the plotted approach is a way of securing and empowering those public groups who are predisposed to be interested in literature, but also to attracting and mobilizing new publics through digital resources, innovative cartographies, workshops, and public excursions. By encouraging potential stakeholders to engage with *Literary Atlas* materials, an accessible entry point has been identified into the assemblage of literature and maps. Plot appears to be useful as a means to sensitize publics to a key facet of the human condition, mobilizing their own person–place relations, and generating interest in different localities, their contextual histories, authors, and relations. Whatever component is accessed (be it the range of maps on the *Literary Atlas* website, for example, the novel itself, the distant or deep map, the plotline spatial narrative, the commissioned artworks, the public exhibitions or workshops), a broader engagement is prompted which encourages a journey into other components of the assembled project, creating an appetite to "feast" on the literary geographies on offer. The evaluation of this engagement has suggested that this is a positive process, generating a circuit of connection which is affective and affirming, and productive of moments of recognition, understanding, challenge, and reflection.

These moments of (self-) recognition, affirmation, and expression may be enjoyable, but a sceptic may suggest that they lack importance as they do not change the world in a material or policy sense. Yet, in this series of inspirations, a new set of possibilities is nonetheless sensed. These moments become fragments, or perhaps even building blocks (echoing the experience of Conran's *Pigeon*), where read-realities become "recognized", in other words, where modern cognition is rebooted to include the felt, sensed, and lived. Through this process, the broader assemblages implicit within a plotted reader's spatial being are concomitantly experienced and appreciated. Such recognition offers opportunities for monetization and commercial activity (through plotted excursions, regional marketing of places as literary and plotted in nature, and associated merchandise), but perhaps more innovatively, also provides opportunities for replacing a sense of geographical identity and community through (in)formal education and cultural literacy. The *Literary Atlas* project has proven that individuals become aware of and interested in a broader set of geographies when prompted to do so by the plotted approach to not only literature but also a range of associated cartographies. Such engagement could be more overtly harnessed through curricula (complementing the Welsh Learned Society "cynefin" approach (2022), for example), and used to focus novels and their plotted reading as a key route into a critical knowledge of particular places, their histories, cultures, literatures, and creative writings, as well as a creative sense of self which is more than modern.

In this way, a plotted approach may not be impactful in a modern sense, but remains directly political, engaging with Solnit's challenging question: "who gets to shape the narrative of our times?" (2019: inside sleeve). The *Literary Atlas*' plotted approach re-politicizes our ontological assumptions about our era and our contribution to it; it thrusts us into taking seriously the conclusion that modern categorizations "about the character of the world . . . [are] an enactment and only partial" (Law, 2015:126). Through doing so, the *Literary Atlas* embraces that which all (plotted) readers of literature are already aware that "we participate in multiple realities" (Law, 2015:126) and live at least part of our lives in borderland spaces. Indeed, Solnit's choice of words gives us further pause for reflection on this fundamental: according to Solnit, our times are shaped by narratives, constructed and formed by positioned and situated ideas; indeed for Solnit, "we live inside [these] ideas" (2019:3). By framing the question and our world in this way, Solnit reminds us that ideas, stories, and novels are not simply make believe but are imbued into the very experience of our everyday. To marginalize these cultural traces from their coingredience in our lives seems to be an act of negligence, and furthermore, to restrict access to these stories and the "foundational" role (after Solnit, 2019: inside sleeve) they play in human identification and community (be it through limiting curricula, closing libraries, or banning books) is overtly political. In this way, the plotted approach to literary geographies can offer not simply a new way to "see" our storied landscapes but to celebrate and harness them too; it offers a tangible opportunity to embrace our read-realities as an integral facet of our lived experience, and use them to foster a critical take on local places and cultures, as a means to generate local economies, and to stimulate creative expression amongst populations. With this in mind, it is perhaps the responsibility of those working on the subject of literary geographies to insist that this read-reality – the borderland space of plot – not simply exists but has value, and is worthy of creative and critical engagement. This may be a "radical" (after Law, 2015) position for literary geographies to pursue, but one that takes seriously its relational commitment. As the *Literary Atlas of Wales and its Borderlands* has demonstrated, the plotted approach is a key means of identifying and embracing (read-)realities as a space of potential commercial opportunity, intellectual and imaginative nourishment, and individual and community well-being. It changes how we consider the power of literature to orientate us and give a sense of connection and relation to ourselves and others in our world.

Note

1 The zombie category is a term coined by Beck and Beck-Gernsheim (2002) as part of a broader individualization and reflexive modernization thesis (see Beck, 1994, 1995). This project posits that the contours of industrial society have begun to dissolve under the dynamism of its rapid success. As a consequence, conventional identity categories (such as family, class, and religion) become less relevant to the positions taken by individuals and groups to define themselves and their behaviour. For Beck, these categories have thus become "zombies": they are living but dead; although their vocabulary is still in popular use, they no longer have any meaningful connection with the post-/late-modern society in which we are living. In other words, these zombie categories obfuscate reality; we no longer know what they stand for, or what lies behind them. In short, they hide more than they illuminate (2002:207). In Beck and Beck-Gernsheim's words: "zombie or living-dead categories . . . blind [us] to the realities and contradictions of globalizing and individualizing modernities" (2002:XXIV).

Bibliography

Abbey, E. (1975) *The Monkey Wrench Gang*. Philadelphia: Lippencott Williams Wilkins.
Abse, D. (1954 / 2022) *Ash on a Young Man's Sleeve*. Cardigan: Parthian.
Ackroyd, P. (2000) *London: The Biography*. London: Chatto and Windus.
Ackroyd, P. (2007) *Thames: Sacred River*. London: Chatto and Windus.
Ackroyd, S. and Hughes, J.A. (1981) *Data Collection in Context*. New York: Longman.
Agnew, J. and Duncan, J. (eds.). (1989) *The Power of Place: Bringing Together Geographical and Sociological Imaginations*. London: Unwin Hyman.
Aitkin, S. (1997) Analysis of texts: Armchair theory and couch potato geography. In Flowerdew, R. and Martin, D. (eds.), *Methods in Human Geography: A Guide for Students Doing a Research Project*. Harlow: Longman, 197–212.
Alexander, N. (2015) On literary geography. *Literary Geographies*, 1, 3–6.
Allington, D. (2011) 'It actually painted a picture of the village and the sea and the bottom of the sea': Reading groups, cultural legitimacy, and description in narrative (with particular reference to John Steinbeck's The Pearl). *Language and Literature*, 20(4), 317–332.
Allington, D. and Swann, J. (2009) Researching literary reading as social practice. *Language and Literature*, 18(3), 219–230.
All Poetry. (n.d.) *Reservoirs*. R.S. Thomas. 1968. https://allpoetry.com/reservoirs. Accessed January 2024.
Amis, K. (1986) *The Old Devils by Kingsley Amis*. London: Vintage.
Anderson, B. (2018) Cultural geography II: The force of representations. *Progress in Human Geography*, 436, 1120–1132.
Anderson, J. (2004) Talking whilst walking: A geographical archaeology of knowledge. *Area*, 36(3), 254–261.
Anderson, J. (2009) Transient convergence and relational sensibility: Beyond the modern constitution of nature. *Emotion, Space, and Society*, 2, 120–127.
Anderson, J. (2012) Relational places: The surfed wave as assemblage and convergence. *Environment & Planning D: Society & Space*, 304, 570–587.
Anderson, J. (2014) *Page and Place: Ongoing Compositions of Plot*. Amsterdam: Rodopi.
Anderson, J. (2015) Mapping the consequences of mobility: Reclaiming jet lag as the state of travel disorientation. *Mobilities*, 101, 1–16.
Anderson, J. (2016a) Between 'distant' and 'deep' digital mapping: Walking the plotlines of Cardiff's literary geographies. In Cooper, D., Donaldson, C. and Murrieta-Flores, P. (eds.), *Literary Mapping in the Digital Age*. London and New York: Routledge, 161–181.
Anderson, J. (2016b) Towards an assemblage approach to literary geography. *Literary Geographies*, 1(2), 120–137.
Anderson, J. (2021) *Understanding Cultural Geography: Place and Traces* (3rd ed.). London: Routledge.
Anderson, J. (2023) *Surfing Spaces*. London: Routledge.
Anderson, J., Adey, P. and Bevan, P. (2010) Positioning place: Polylogic approaches to research methodology. *Qualitative Research*, 10(5), 589–604.

Anderson, J. and Erskine, K. (2014) Tropophilia: A study of people, place and lifestyle travel. *Mobilities,* 91, 130–145.

Anderson, J. and Jones, K. (2009) The difference that place makes to methodology: Uncovering the 'lived space' of young people's spatial practices. *Children's Geographies*, 7(3), 291–303.

Anderson, J. and Peters, K. (2014a) 'A perfect and absolute blank': Human geographies of water worlds. In Anderson, J. and Peters, K. (eds.), *Water Worlds: Human Geographies of the Ocean*. Farnham: Ashgate, 103–118.

Anderson, J. and Peters, K. (eds.). (2014b) *Water Worlds: Human Geographies of the Ocean.* Farnham: Ashgate.

Anderson, J. and Saunders, A. (2016) Relational literary geographies: Co-producing page and place. Theme editorial. *Literary Geographies*, 1(2), 115–119.

Anderson, J. and Stanley, K. (2019) Attuning to the affective in literary tourism: Emotional states in Aberystwyth, Mon Amour. *Tourism Geographies*, 24(2–3), 435–456.

Anderson, K. (2010) From 'zombies' to 'coyotes': Environmentalism where we are. *Environmental Politics*, 19(6), 973–991.

Arskey, H. and Knight, P. (1999) *Interviewing for Social Scientists: An Introductory Resource with Examples*. London: Sage.

Azzopardi, T. (2000) *The Hiding Place*. London: Picador.

Badiou, A. (2001) *Ethics: An Essay on the Understanding of Evil*. London: Verso.

Badiou, A. (2005) *Being and Event.* London: Continuum.

Badiou, A. (2006) The Paris commune: A political declaration on politics. In Badiou, A. (ed.), *Polemics.* London: Verso, 257–290.

Bakhtin, M. (1984) *Rabelais and His World.* Bloomington: Indiana University Press.

Barnes, A. (2013) Geo/graphic design: The liminal space of the page. *The Geographical Review*, 103(2), 164–176.

Barnes, A. (2018) *Creative Representations of Place.* London: Routledge.

Barry, A., Born, G. and Weszkalnys, G. (2008) Logics of interdisciplinarity. *Economy and Society*, 37, 20–49.

Barthes, R. (1977) *Mythologies*. London: Paladin.

Bassett, K. (2008) Thinking the event: Badiou's philosophy of the event and the example of the Paris Commune. *Environment & Planning D: Society & Space*, 26, 895–910.

Bauman, Z. (2000) *Liquid Modernity*. Cambridge: Polity Press.

BBC (2012) Mid Wales Books. *Aber Mon Amour*. https://downloads.bbc.co.uk/wales/archive/bbc-mid-wales-books-malcolm-pryce-interview.pdf. Accessed January 2024.

Becher, T. (1989) *Academic Tribes and Territories: Intellectual Enquiry and the Cultures of Disciplines.* Milton Keynes, England: Society for Research into Higher Education/Open University Press.

Beck, U. (1992) *Risk Society: Towards a New Modernity*. London: Sage.

Beck, U. (1994) The reinvention of politics: Towards a theory of reflexive modernisation. In Beck, U., Giddens, A. and Lash, S. (eds.), *Reflexive Modernization: Politics, Tradition and Aesthetics in the Modern Social Order*. Cambridge: Polity Press, 1–55.

Beck, U. (1995) *Ecological Politics in an Age of Risk.* Cambridge: Polity Press.

Beck, U. and Beck-Gernsheim, E. (1995) *The Normal Chaos of Love*. Cambridge: Polity Press.

Beck, U. and Beck-Gernsheim, E. (2002) *Individualization*. London: Sage.

Benedikt, M. (1991) Cyberspace: Some proposals. In Benedikt, M. (ed.), *Cyberspace: First Steps*. Cambridge, MA: MIT Press.

Bennett, A. and Royle, N. (2016) *An Introduction to Literature, Criticism, and Theory* (5th ed.). London and New York: Routledge.

Benwell, B. (2012) Commonsense anti-racism in book group talk: The role of reported speech. *Discourse and Society,* 23(4), 359–376.

Berry, R. (1970) *So Long, Hector Bebb*. Cardigan: Parthian.

Binnie, J., Edensor, T., Holloway, J., Millington, S. and Young, C. (2007) Mundane mobilities, banal travels. *Social & Cultural Geography*, 8(2), 165–174.

Boakye, J. (2023) The big idea: Do we need to dismantle the literary canon? *The Guardian*. www.theguardian.com/books/2023/jun/12/the-big-idea-do-we-need-to-dismantle-the-literary-canon. Accessed June 2024.

Bodenhamer, D. (2021) The variety of deep maps. In Bodenhamer, D., Corrigan, J. and Harris, T. (eds.), *Making Deep Maps*. London: Routledge, 1–15.

Bodenhamer, D., Corrigan, J. and Harris, T. (eds.). (2015) *Deep Maps and Spatial Narratives*. Bloomington, IN: Indiana University Press.

Bondi, L. (1990) Feminism, postmodernism and geography: Space for women? *Antipode*, 22(2), 156–167.

Bondi, L. (1998) Sexing the city. In Fincher, R. and Jacobs, J.M. (eds.), *Cities of Difference*. New York: Guilford Press, 177–200.

Bondi, L. (2005a) Making connections and thinking through emotions: Between geography and psychotherapy. *Transactions of the Institute of British Geographers*, 30, 433–448.

Bondi, L. (2005b) The place of emotions in research: From partitioning emotion and reason to the emotional dynamics of research relationships. In Davidson, J., Bondi, L. and Smith, M. (eds.), *Emotional Geographies*. Aldershot: Ashgate, 231–246.

Bondi, L., Davidson, J. and Smith, M. (2005) Introduction: Geography's 'emotional turn'. In Davidson, J., Bondi, L. and Smith, M. (eds.), *Emotional Geographies*. Aldershot: Ashgate, 1–18.

Bondi, L. and Smith, M. (2005) *Emotional Geographies*. Aldershot: Ashgate.

Bonnett, A. (2009) The dilemmas of radical nostalgia in British psychogeography. *Theory, Culture, Society*, 25(1), 45–70.

Bourdieu, P. (1977) *Outline of a Theory of Practice*. Cambridge: Cambridge University Press.

Bourdieu, P. (1991) *Language and Symbolic Order*. Cambridge: Polity Press.

Bradbury, R. (1953/2008) *Fahrenheit 451*. London: HarperCollins.

Bradshaw, A. and Brown, S. (2018) Up rising: Rehabilitating J.G. Ballard's high-rise with R.D. Laing and Lauren Berlant. *Environment and Planning D: Society and Space*, 36(2), 331–349.

Braun, B. (2004) Nature and culture: On the career of a false problem. In Duncan, J., Johnson, N. and Schein, R. (eds.), *A Companion to Cultural Geography*. Malden: Blackwell, 151–179.

Bridge, G. (2018) The map is not the territory: A sympathetic critique of energy research's spatial turn. *Energy Research and Social Science*, 36, 11–20.

British Association for Modernist Studies. (n.d.) Modernist studies. https://bams.ac.uk/. Accessed June 2024.

British Library. (2012a) Writing Britain: Literature and landscape. www.bl.uk/writingbritain. Accessed January 2013.

British Library. (2012b) Introduction to "Writing Britain: Wastelands to Wonderlands". https://www.youtube.com/watch?v=yrCCF3IMN2k&t=11s. Accessed January 2023. British Library. (n.d.) Literature and land. www.bl.uk/onlinegallery/onlineex/literland/index.html. Accessed January 2013.

Brosseau, M. (1994) Geography's literature. *Progress in Human Geography*, 18(3), 333–353.

Brosseau, M. (2017) In, of, out, with, and through: New perspectives in literary geography. In Tally, R.T., Jr. (ed.), *The Routledge Handbook of Literature and Space*. New York & London: Routledge, 9–27.

Brown, C. (2018) Spacing events: Charting choreo- spatial dramaturgies. In Duxbury, N., Garret-Petts, W. and Longley, A. (eds.), *Artistic Approaches to Cultural Mapping*. London and New York: Routledge.

Brown, K. (2022) Art markets, epistemic authority, and the institutional curation of knowledge. *Cultural Studies*, 37, 626–646. DOI: 10.1080/09502386.2022.2030777.
Brown, M. and Humberstone, B. (eds.). (2015) Seascapes: Shaped by the sea. In *Embodied Narratives and Fluid Geographies*. Farnham: Ashgate.
Brown, P. and Irwin, M. (eds.). (2008) *Literature & Place, 1800–2000*. Oxford: Peter Lang.
Burton, J. (2023) When I am anxious or sad I reach for Hilary Mantel's Wolf Hall. *The Guardian*. www.theguardian.com/books/2023/may/19/jessie-burton-when-i-am-anxious-or-sad-i-reach-for-hilary-mantels-wolf-hall. Accessed May 2023.
Bush, D. (1995) *The Genre of Silence*. Bridgend: Seren.
Bushell, S. (2012) The slipperiness of literary maps: Critical cartography and literary cartography. *Cartographica*, 47, 149–160.
Callon, M. (1986) Some elements in a sociology of translation. In Law, J. (ed.), *Power, Action, Belief*. London: Routledge, 19–34.
Cameron, D. (2006) Performing gender identity: Young men's talk and the construction of heterosexual masculinity [1997]. In *On Language and Sexual Politics*. Abingdon, Oxon: Routledge, 61–74.
Cameron, E. (2012) New geographies of story and storytelling. *Progress in Human Geography*, 36(5), 573–592.
Caquard, S. (2011) Cartography I: Mapping narrative cartography. *Progress in Human Geography*, 37, 135–144.
Caquard, S. and Bryne, A. (2009) Mapping globalization: A conversation between a filmmaker and a cartographer. *The Cartographic Journal*, 46(4), 372–378.
Carson, R. (1962) *Silent Spring*. Harmondsworth: Penguin.
Casey, E. (2001) Between geography and philosophy: What does it mean to be in the place-world? *Annals of the Association of American Geographers*, 914, 683–693.
Castán Broto, V. and Baker, L. (2018) Introduction. Special issue 'spatial adventures in energy: Emerging geographies of energy production and use'. *Energy Research & Social Science*, 36, 1–10.
Castree, N. and MacMillan, T. (2002) Dissolving dualisms: Actor networks and the reimagination of nature. In Castree, N. and Braun, B. (eds.), *Social Nature*. Oxford: Blackwell, 208–224.
Chaabane, B. (2020) A carnivalistic reading of *The Wedding of Zein* by Taib Salih. *The Journal of North African Studies*. DOI: 10.1080/13629387.2020.1838903.
Charlton, C. (2014) The weight of curious space: Rhetorical events, hackerspace, and emergent multimodal assessment. *Computers and Composition*, 31, 29–42.
Chatwin, B. (1982) *On the Black Hill*. London: Vintage.
Chatwin, B. (1987) *Songlines*. London: Vintage.
Chavinier, E., Lanoix, C., Levy, J. and Mauron, V. (2016) What the atlas does to the map. In Levy, J. (ed.), *A Cartographic Turn: Mapping and the Spatial Challenge in Social Sciences*. Lausanne: Routledge/EPFL Press.
Childress, C. and Friedkin, N. (2012) Cultural reception and production: The social construction of meaning in book clubs. *American Sociological Review*, 77(1), 45–68.
Conan Doyle, A. (1903) *The Works of Sir Arthur Conan Doyle*. London: John Murray.
Conradson, D. (1997) Focus Groups. In Flowerdew, R. and Martin, D. (eds.), *Methods in Human Geography: A Guide for Students Doing a Research Project*. Harlow: Longman.
Conran, A. (2016a) *Pigeon*. Cardigan: Parthian.
Conran, A. (2016b) Pigeon. www.fromthemargins.co.uk/pages/reviews_all/Alys_Conran_Pigeon.html. Accessed June 2016.
Cook, I. (1997) Participant observation. In Flowerdew, R. and Martin, D. (eds.), *Methods in Human Geography: A Guide for Students Doing a Research Project*. Harlow: Longman, 127–150.
Cook, I., Crouch, D., Naylor, S. and Ryan, J. (eds.). (2000) *Cultural Turns/Geographical Turns*. Harlow: Prentice Hall.

Cooper, D., Donaldson, C. and Murrieta-Flores, P. (eds.). (2016) *Literary Mapping in the Digital Age*. Farnham: Ashgate.

Country, B., Wright, S., Suchet-Pearson, S., Lloyd, K., Burarrwanga, L., Ganambarr, R., Ganambarr-Stubbs, M., Ganambarr, B. and Maymuru, D. (2015) Working with and learning from Country: Decentring human authority. *Cultural Geographies*, 22(2), 269–283.

Courage, C. (2019) *Arts in Place: The Arts, the Urban, and Social Practice*. London: Routledge.

Cresswell, T. (1996) *In Place/Out of Place*. Minneapolis: University of Minnesota Press.

Cresswell, T. (2004) *Place: A Short Introduction*. Oxford: Blackwell.

Cresswell, T. (2019) *Maxwell Street: Writing and Thinking Place*. Chicago: University of Chicago Press.

Dafydd, F. (2008) *Twenty Thousand Saints*. Ceredigion: Alcemie.

Dahl, R. (1970) *Fantastic Mr Fox*. London: Allen and Unwin.

Dahl, R. (1988) *Mathilda*. Harmondsworth: Penguin.

Daniels, S., Delyser, D., Entrikin, N. and Richardson, D. (eds.). (2011) *Envisioning Landscapes, Making Worlds*. London & New York: Routledge.

Davies, R. (1928) *The Withered Root*. Cardigan: Parthian.

Davies, S. (2013) *Awakening*. Cardigan: Parthian.

de Certeau, M. (1984) *The Practice of Everyday Life*. Berkeley: University of California Press.

DeGraff, A. (2015) *Plotted: A Literary Atlas*. San Francisco: Pulp.

Delanda, M. (2006) *A New Philosophy of Society: Assemblage Theory and Social Complexity*. London: Continuum.

Deleuze, G. (1985) Nomad thought. In Alison, D. (ed.), *The New Nietzsche*. Cambridge, MA: MIT Press, 142–149.

Deleuze, G. and Guattari, F. (2003) *A Thousand Plateaux: Capitalism & Schizphrenia*. Minneapolis: University of Minnesota Press.

Delyser, D. and Hawkins, H. (2014) Introduction: Writing creatively – Process, practice, and product. *Cultural Geographies*, 21(1), 131–134.

Denison, J. and Markula, P. (2003) *Moving Writing: Crafting Movement in Sports Research*. New York: Peter Lang.

Denzin, N. (1998) The new ethnography. *Journal of Contemporary Ethnography*, 27(3), 405–415.

DePalma, M.J. and Alexander, K.P. (2015) A bag full of snakes: Negotiating the challenges of multimodal composition. *Computers and Composition*, 37, 182–200.

Derrida, J. (1989) Jacques Derrida interview by Christopher Norris. In Papadakis, A., Cooke, C. and Benjamin, A. (eds.), *Deconstruction: Omnibus Volume*. London: Academy Editions, 71–78.

Dewsbury, J.-D. (2011) The Deleuze-Guattarian assemblage: Plastic habits. *Area*, 43(2), 148–153.

Dickens, T. and Dickens, L. (2022) Dreamlands: Stories of enchantment and excess in a search for lost sensations. *Cultural Geographies*, 29(1), 23–43.

Dillwyn, A. (1880) *The Rebecca Rioter*. Aberystwyth: Honno.

Dodge, M., Kitchin, R. and Perkins, C. (2009) Thinking about M. In Dodge, M., Kitchin, R. and Perskins, C. (eds.), *Rethinking Maps*. London: Routledge, 1–25.

Doel, M. (1999) *Poststructuralist Geographies: The Diabolical Art of Spatial Science*. Lanham, MD: Rowman & Littlefield.

Doel, M. (2018) Literary space uncut. *Literary Geographies*, 4(1), 42–49.

Doel, M. (2023) The unhinged hinge: Pegged out, pinned down, and folded away. *Literary Geographies*, 9(2), 279–284.

Donaldson, A., Ward, N. and Bradley, S. (2010) Mess among disciplines: Interdisciplinarity in environmental research. *Environment and Planning A*, 42, 1521–1536.

Driver, F. (2000) *Geography Militant: Cultures of Exploration and Empire*. Oxford: Blackwell.

Drucker, J. (2008) The virtual codex from page space to E-space. In Schreibman, S. and Siemans, R. (eds.), *A Companion to Digital Literary Studies*. Oxford: Blackwell, 333–342.

Duchamp, M. (1973) *The Writings of Marcel Duchamp*, eds. Michel Sanouillet and Elmer Peterson. New York: Da Capo.

Dunthorne, J. (2008) *Submarine*. London: Hamish Hamilton.

Duxbury, N., Garrett-Petts, W.F. and Longley, A. (eds.). (2018) *Artistic Approaches to Cultural Mapping: Activating Imaginaries and Means of Knowing*. London: Routledge.

Eanes, F.R., Silbernagel, J., Robinson, P. and Hart, D. (2020) Spatial narratives in theory and practice for landscape conservation and public engagement. *Landscape Journal*, 38(1), 1–2.

Eco, U. (1986) *Travels in Hyperreality*. San Francisco: Harcourt.

Eisteddfod. (2022) Gold medal for fine art. https://eisteddfod.wales/node/1506. Accessed July 2022.

Eliot, T.S. (1943) Little gidding. In *Four Quartets*. Gardners Books; Main edition, April 30, 2001.

Evans, R. (2013) A Welsh landscape. http://thecuriousastronomer.wordpress.com/2013/02/21/a-welsh-landscape-r-s-thomas-poem/. Accessed October 2013.

Eyles, J. (1988) Interpreting the geographical world: Qualitative approaches in geographical research. In Eyles, J. and Smith, D.M. (eds.), *Qualitative Methods in Human Geography*. Cambridge: Polity Press, 1–16.

Felgenhauer, T. and Gabler, K. (eds.). (2018) *Geographies of Digital Culture*. London: Routledge.

Ferguson, B. (1996) Exhibition rhetorics material speech and utter sense. In Greenberg, R., Ferguson, B. and Nairne, S. (eds.), *Thinking About Exhibitions*. Abingdon: Routledge, 175–190.

Finch, P. (1987) *Selected Poems*. Bridgend: Poetry Wales Press.

Fletcher, L. (2021) *Life Force: A Painter's Response to the Nature Poetry of Ted Hughes*. Llanelli: Bird Eye Books/Graffeg.

Folke Henningsen, A., Gregersen, A. and Vest Hansen, M. (2020) Introduction: Thinking and doing exhibitions. In Folke Henningsen, A., Gregersen, A. and Vest Hansen, M. (eds.), *Curatorial Challenges: Interdisciplinary Perspectives on Contemporary Curating*. London: Routledge, 1–5.

Foucault, M. (1973) *The Order of Things*. New York: Vintage/Random House.

Foucault, M. (1980) *Power/Knowledge*. New York: Pantheon.

Foucault, M. (1984) *History of Sexuality*, Vol. 1. Harmondsworth: Penguin.

Foucault, M. (1995) *Discipline and Punish: The Birth of the Prison*. New York: Doubleday.

Frank, J. (1945) Spatial form in modern literature: An essay in two parts. *The Sewanee Review*, 53(2), 221–240.

Fuller, D. (2008) Reading as social practice: The beyond the book project. *Popular Narrative Media*, 1(2), 211–216.

Fuller, D. and Procter, J. (2009) Reading as "social glue"? *Moving Worlds*, 9(2), 26–40.

Fuller, D. and Rehberg Sedo, D. (2013) *Reading Beyond the Book: The Social Practices of Contemporary Literary Culture*. London: Routledge.

Gallie, M. (1959/2011 edition) *Strike for a Kingdom*. Honno: Aberystwyth.

Game, A. (1997) Sociology's emotions. *CRSA/RSCA*, 34(4), 385–399.

Game, A. and Metcalfe, A. (1996) *Passionate Sociology*. London: Sage.

Gao. (2020) Curation as a method of cultural inquiry. *Frontiers in Art Research*, 2(7). https://francis-press.com/papers/2875. Accessed December 23, 2022.

Garner, A. (1967/2002) *The Owl Service*. London: Collins.

Garner, A. (2010) *The Voice That Thunders*. London: Harvill Press.

Gill, N., Conlon, D., Moran, D. and Burridge, A. (2018) Carceral circuitry: New directions in carceral geography. *Progress in Human Geography*, 42(2), 183–204.

Goldberg, N. (1991) *Wild Mind: Living the Writer's Life*. London: Rider.

Gramich, E. (2019) A dancing clod of earth: Review of 'cartographic imaginaries'. *Planet Magazine.* www.planetmagazine.org.uk/planet-online/234/eluned-gramich. Accessed July 2019.

Gray, C. (2010) Museums, galleries, politics and management. *Public Policy and Administration*, 26(1), 45–61. https://journals.sagepub.com/doi/abs/10.1177/0952076710365436?casa_token=LcWTV5NrVPoAAAAA:7w70gA8IFlE6fhcQN0pa28fVPVRAk-eg3vBy_Sm1W0yQfwUyVq-mo-VyzdkLFgw7zGFLX5_deDi1. Accessed October 2022.

Greenspan, B. (2011) Songlines in the streets: Story mapping with itinerant hypernarrative. In Thomas, B. and Page, R. (eds.), *New Narratives Stories and Storytelling in the Digital Age*. Lincoln: University of Nebraska Press, 153–169.

Gregory, I. and Cooper, D. (2013) Geographical technologies and the interdisciplinary study of peoples and cultures of the past. *Journal of Victorian Culture*, 18(2), 265–272.

Gregory, I. and Donaldson, C. (2016) Geographical textual analysis. In Cooper, D., Donaldson, C. and Murrieta-Flores, P. (eds.), *Rethinking Literary Mapping*. London: Routledge, 67–87.

Griffiths, N. (2001) *Sheepshagger.* London: Vintage.

Griswold, W., Lenaghan, E. and Naffziger, M. (2011) Readers as audiences. In Virginia Nightingale (ed.), *The Handbook of Media Audiences*. Hoboken: Blackwell Publishing, 19–41.

Griswold, W., McDonnell, T. and Wright, N. (2005) Reading and the reading class in the twenty-first century. *Annual Review of Sociology*, 31, 127–141.

Guillory, J. (2000) The ethical practice of modernity. In Garber, M., Hanssen, B. and Walkowitz R. (eds.), *The Turn to Ethics*. London: Routledge, 29–46.

Hakim, C. (1987) *Research Design: Strategies and Choices in the Design of Social Research*. London: Unwin Hyman.

Hall, S. (2004) I, Mercator. In Harmon, K. (ed.), *You Are Here: Personal Geographies and Other Maps of the Imagination*. New York: Princeton Architectural Press, 15–19.

Hamlin, M. and Speer, J. (2018) The politics of conceptualizing the carceral: A commentary on Moran et al. (2017). *Progress in Human Geography*, 42(5), 799–802.

Haraway, D. (1988) Situated knowledges: The science question in feminism and the privilege of partial perspective. *Feminist Studies*, 14(3), 575–599.

Haraway, D. (2003) *The Companion Species Manifesto*. Chicago: University of Chicago Press.

Harley, J.B. (1988) Maps, knowledge and power. In Cosgrove, D. and Daniels, S. (eds.), *The Iconography of Landscape: Essays on the Symbolic Representation, Design and Use of Past Environments*. London: Routledge, 277–312.

Harmon, K. (2003) *You Are Here: Personal Geographies and Other Maps of the Imagination*. Princeton: Princeton Architectural Press.

Harris, T. (2015) Deep geography – Deep mapping. In Bodenhamer, D., Corrigan, J. and Harris, T. (eds.), *Deep Maps and Spatial Narratives*. Bloomington, IN: Indiana University Press, 28–53.

Hartley, J. (2001) *Reading Groups*. Oxford: Oxford University Press.

Harvey, D. (1996) *Science, Nature and the Geography of Difference*. Oxford: Blackwell.

Hawkins, D. (2011) Dialogues and doings: Sketching the relationships between geography and art. *Geography Compass*, 5(7), 464–478.

Hawkins, H. (2012) Geography and art. An expanding field: Site, the body and practice. *Progress in Human Geography*, 37(1), 52–71.

Hawkins, H. (2015) Creative geographic methods: Knowing, representing, intervening. On composing place and page. *Cultural Geographies*, 22(2), 247–268.

Hawkins, H. (2021) *Geography, Art, Research: Artistic Research in the GeoHumanities*. London: Routledge.

Hennessy, C. and Means, R. (2015) Meeting the challenge of interdisciplinarity: Lessons and issues from the Grey and Pleasant Land project on rural ageing. *Journal of Geography in Higher Education*, 39(2), 195–205.

Herbert, D. (2001) Literary places, tourism and the heritage experience. *Annals of Tourism Research*, 28(2), 312–333.

Hermon, P. (2006) *Hillwalking in Wales: Volume 2*. Kendal: Ciceron Press.

Hintz, C. (2020) Picture books. https://picturebookbuilders.com/2020/02/carol-hintz-talks-about-picture-book-pagination/?doing_wp_cron=1689863327.4126980304718017578125. Accessed January 2024.

Ho, E. (2006) Postimperial landscapes: "Psychogeography" and Englishness in Alan Moore's graphic novel from hell: A melodrama in sixteen parts. *Cultural Critique*, 63, 99–121.

Hoare, J. (2019) Heart on your sleeve? Emotion, wearable technology and digital heritage. In *2018 3rd Digital Heritage International Congress (DigitalHERITAGE) Held Jointly with 2018 24th International Conference on Virtual Systems & Multimedia (VSMM 2018)*, San Francisco, CA, 1–4.

Holton, M. and Riley, M. (2014) Talking on the move: Place- based interviewing with undergraduate students. *Area*, 46(1), 59–65.

Hones, S. (2008) Text as it happens: Literary geography. *Geography Compass*, 2(5), 1301–1317.

Hones, S. (2014) *Literary Geographies: Narrative Space in Let the Great World Spin*. New York: Palgrave Macmillan.

Hones, S. (2018) Literary geography and spatial literary studies. *Literary Geographies*, 4(2), 146–149.

Hones, S. (2022a) Interspatiality. *Literary Geographies*, 8(1), 15–18.

Hones, S. (2022b) *Literary Geography*. London and New York: Routledge.

Hooks, B. (1991) *Yearning: Race, Gender and Cultural Politics*. New York: Turnaround Press.

Hoppen, A., Brown, L. and Fyall, A. (2014) Literary tourism: Opportunities and challenges for the marketing and branding of destinations? *Journal of Destination Marketing and Management*, 3, 37–47.

Hornsey, R. (2012) Listening to the Tube Map: Rhythm and the historiography of urban map use. *Environment and Planning D: Society and Space*, 30, 675–693.

Hughes, T. (2008) *Revenant*. London: Picador.

Humphreys, E. (1958) *A Toy Epic*. Bridgend: Seren.

Hurni, L. and Piatti, B. (2019) Mapping and analysing the geography of fiction with interactive tools. *European Atlas*. https://karto.ethz.ch/en/research/completed-projects/a-literary-atlas-of-europe.html

Idhe, D. (1990) *Technology and the Lifeworld*. Bloomington: Indiana University Press.

Ingold, T. (2007) *Lines: A Brief History*. London and New York: Routledge.

Jarvis, R. (1997) *Romantic Writing and Pedestrian Travel*. Basingstoke: Macmillan.

Johnson, A.K. (2019) Co-curation, re-framing cultural spaces from an outsider lens. In *Gender – Diversity – Intersectionality (New) Perspectives in Adult Education*. Münster Waxmann, 191–195.

Johnson, L. (2017) Teaching intelligence: Briefing books, murder boards, and stirring scenarios. *Intelligence and National Security*, 32(7), 961–971.

Johnson, N. (2004) Fictional journeys: Paper landscapes, tourist trails and Dublin's literary texts. *Social & Cultural Geography*, 5(1), 91–107.

Johnston, R. (2000) Intellectual respectability and disciplinary transformation? Radical geography and the institutionalisation of geography in the USA since 1945. *Environment and Planning A*, 32, 971–990.

Jones, E. (2018) What literature is spatial? *Literary Geographies*, 4(1), 38–41.

Jones, G. (1956) *The Valley, The City, The Village*. Cardigan: Parthian.

Jones, L. (1937) *Cwmardy and We Live*. Cardigan: Parthian.

Jones, L. (2004) *Mr Vogel*. Bridgend: Seren.

Jones, M. (2009) Phase space: Geography, relational thinking, and beyond. *Progress in Human Geography*, 33(4), 487–506.

Katz, C. (1994) Playing the field: Questions of fieldwork in geography. *The Professional Geographer*, 46(1), 67–72.

Kirshenblatt-Gimblett, B. (2000) Museum as catalyst. In Keynote address presented at the Museums 2000: Confirmation or Challenge conference, organized by ICOM Sweden, the Swedish Museum Association and the Swedish Travelling Exhibition/Riksutställningar in Vadstena, 1–19.

Kitchin, R. (2016) Foreword. In Levy, J. (ed.), *A Cartographic Turn*. Lausanne: EPFL Press, 5–7.

Kitchin, R. and Dodge, M. (2007) Rethinking maps. *Progress in Human Geography*, 31, 331–344.

Kitchin, R., Dodge, M. and Perkins, C. (2011a) Introductory essay: Conceptualising mapping. In Kitchin, R., Dodge, M. and Perkins, C. (eds.), *The Map Reader Theories of Mapping Practice and Cartographic Representation*. Oxford: Wiley Blackwell, 2–7.

Kitchin, R., Dodge, M. and Perkins, C. (eds.). (2011b) *The Map Reader*. Oxford: Wiley.

Kitchin, R., Perkins, C. and Dodge, M. (2009) Thinking about Maps. In Dodge, M., Kitchin, R. and Perkins, C. (eds.), *Rethinking Maps: New Frontiers of Cartographic Theory*. Abingdon: Routledge, 1–25.

Klein, J.T. (2010) A taxonomy of interdisciplinarity. In Frodemann, R., Klein, J.T. and Mitcham, C. (eds.), *The Oxford Handbook of Interdisciplinarity*. Oxford: Oxford University Press, 15–30.

Kuhn, T. (1970) *The Structure of Scientific Revolutions* (2nd ed.). Chicago, IL: University of Chicago Press.

Kuzmičová, A. and Cremin, T. (2021) Different fiction genres take children's memories to different places. *Cambridge Journal of Education*. DOI: 10.1080/0305764X.2021.1937523.

Laclau, E. (2004) An ethics of militant engagement. In Hallward, P. (ed.), *Think Again. Alain Badiou and the Future of Philosophy*. London & New York: Continuum, 120–137.

Laffer, A. (2021) When readers talk about characters as if they were real, how do they talk about them? Empathy and gossip in reading group discourse. *Poetics*, 85, 101503.

Laja, P. (2022) 10 Useful findings about how people view websites. https://cxl.com/blog/10-useful-findings-about-how-people-view-websites/. Accessed January 2024.

Lakiri Dutt, K. and Samanta, G. (2013) *Dancing With the River*. New Haven: Yale University Press.

Larsen, R. (2018) Connecting contours. In Lewis-Jones, H. (ed.), *The Writer's Map*. London: Thames Hudson.

Latour, B. (1993) *We Have Never Been Modern*. Cambridge: Harvard University Press.

Laurier, E. and Philo, C. (1999) X-morphising: Review essay of Bruno Latour's *Aramis, or the Love of Technology*. *Environment and Planning A*, 31, 1047–1071.

Law, J. (2015) What's wrong with a one-world world? *Distinktion: Scandinavian Journal of Social Theory*, 16(1), 126–139.

Leahy, H. (2012) *Museum Bodies: The Politics and Practices of Visiting and Viewing*. Farnham: Ashgate Publishing Limited.

Learned Society. (2022) Wales and the world: Cynefin, colonialism and global interconnections. www.learnedsociety.wales/wales-and-the-world-cynefin-colonialism-and-global-interconnections/. Accessed January 2024.

Learned Society. (n.d.) Wales studies research snapshot. www.learnedsociety.wales/our-publications/wales-studies-research-snapshot/. Accessed January 2024.

Least Heat-Moon, W. (1999) *PrairyErth (a Deep Map)*. Boston and New York: Mariner Books.

Lewis, C. (1960) *Studies in Words*. Cambridge: Cambridge University Press.

Lewis-Jones, H. (2018) *The Writers Map: An Atlas of Imaginary Lands*. London: Thames and Hudson.

Leypoldt, G. (2021) Spatial reading: Evaluative frameworks and the making of literary authority. *American Journal of Cultural Sociology*, 9, 150–176.

Lilley, K. (2011) Digital cartographies and medieval geographies. In Daniels, S., Delyser, D., Entrikin, N. and Richardson, D. (eds.), *Envisioning Landscapes, Making Worlds*. London & New York: Routledge, 25–33.

Lisahunter. (2019) *Surfing, Sex, Genders and Sexualities*. London and New York: Routledge.

Literature Wales. (2024) About Us. https://www.literaturewales.org/. Accessed January 2024.

The Londonist. (2022) Plaques to people who never existed. https://londonist.com/london/secret/london-plaques-to-people-who-never-existed. Accessed January 2024.

Long, E. (1986) Women, reading, and cultural authority: Some implications of the audience perspective in cultural studies. *American Quarterly*, 38(4), 591–612.

Long, E. (1987) Reading groups and the postmodern crisis of cultural authority. *Cultural Studies*, 1(3), 306–327.

Long, E. (1992) Textual interpretation as collective action. *Discourse: Journal for Theoretical Studies in Media and Culture*, 14(3), 104–130.

Long, E. (2003) *Book Clubs: Women and the Uses of Reading in Everyday Life*. Chicago: University of Chicago Press.

Lossau, J. and Lippuner, R. (2004) Geography and the spatial turn. *Erdkunde*, 58(3), 201–211.

Lyotard, J.L. (1990) *Duchamp's Trans/formers*, trans. Ian McLeod. Venice, California: Lapis Press.

MacFarlane, R. (2013) *The Old Ways*. London: Penguin.

Machen, A. (1894) *The Great God Pan*. London: John Lane.

Malkki, L. (1992) National geographic: The rooting of peoples and the territorialization of national identity among scholars and refugees. *Cultural Anthropology*, 7, 24–44.

Mansfield, C. (2015) *Researching Literary Tourism*. Bideford: Shadows Books & Media.

Martin, A. (2020) *Reacher Said Nothing. Lee Child and the Making of Make Me.* Cambridge: Polity.

Martin, R. (1999) Critical survey: The new "geographical turn" in economics: Some critical reflections. *Cambridge Journal of Economics*, 23, 65–91.

Martinez, M.A. and Herman, L. (2020) Real readers reading Wasco's 'City': A storyworld possible selves approach. *Language and Literature*, 29(2), 147–170.

Massey, D. (1994) *Space, Place and Gender*. Cambridge: Polity Press.

Massey, D. (2005) *For Space*. London: Sage.

Massey, D. (2007) *World City*. Cambridge: Polity Press.

McDowell, L. (1999) *Gender, Place and Identity: Understanding Feminist Geographies*. Minneapolis: University of Minnesota Press.

McLaughlin, D. (2018) Teaching literary geographies in British classrooms. *Literary Geographies*, 4(1), 57–61.

McLaughlin, D. (2023) Mapping enchanted landscapes in Philip Weller's *The Dartmoor of* The Hound of the Baskervilles. *Cultural Geographies*, 30(2), 259–277.

McLucas, C. (2014) Deep mapping. https://cliffordmclucas.info/deep-mapping.html. Accessed June 2024.

Meredith, C. (1988) *Shifts*. Parthian: Cardigan.

Merleau-Ponty, M. (1962) *Phenomenology of Perception*. London: Routledge & Kegan Paul.

Miller, T. (2013) Reviving political aesthetics (after Duchamp, even) 2013 1 1 71–104 affirmations of the modern. https://affirmationsmodern.com/articles/78/galley/154/download/

Mitchell, P. (2011) 'The stratified record upon which we set our feet': The spatial turn and the multilayering of history, geography, and geology. In Dear, M., Ketchum, J., Luria, S. and Richardson, D. (eds.), *GeoHumanities: Art, History, Text at the Edge of Place*. Abingdon: Routledge, 71–83.

Monmonier, M. (2007) Cartography: The multidisciplinary pluralism of cartographic art, geospatial technology, and empirical scholarship. *Progress in Human Geography*, 31(3), 371–379.

Moorcock, M. (1988) *Mother London*. London: Secker and Warburg.
Moran, J. (2022) A virtual island journey: Place and place writing in lockdown. *Cultural Geographies*, 29(1), 115–128.
Moretti, F. (1998) *Atlas of the European Novel, 1800–1900*. London: Verso.
Moretti, F. (2005) *Graphs, Maps, Trees: Abstract Models for a Literary History*. London: Verso.
Morris, J. (2009) *Afterword Prichard, C. One Moonlit Night*. Edinburgh: Canogate.
Müller, D. (2006) Unplanned development of literary tourism in two municipalities in rural Sweden. *Scandinavian Journal of Hospitality and Tourism*, 6(3), 214–228.
Murdoch, J. (2005) *Post-Structural Geography: A Guide to Relational Space*. Thousand Oaks, CA: Sage.
Naramore Maher, S. (2014) *Deep Map Country: Literary Cartography of the Great Plains*. Lincoln & London: University of Nebraska Press.
Nicholls, C. (2019) A wild roguery: Bruce Chatwin's songlines reconsidered. *Text Matters*, 9, 22–49.
Nissani, M. (1997) Ten cheers for interdisciplinarity: The case for interdisciplinary knowledge and research. *The Social Science Journal*, 34(2), 201–216.
Noxolo, P., Raghuram, P. and Madge, C. (2008) 'Geography is pregnant' and 'geography's milk is flowing': Metaphors for a postcolonial discipline? *Environment and Planning D: Society and Space*, 26, 146–168.
Nygaard, M. (2018) Bakhtinian Carnivalesque and Paul's Foolish and Scandalous Gospel. *Biblical Interpretation*, 26, 369–390.
Ong, W.J. (1971) World as view and world as event. In Shepard, P. and McKinley, D. (eds.), *Environ Mental: Essays on the Planet as a Home*. Boston: Houghton Mifflin, 61–79.
Orwell, G. (1949) *1984*. London: Secker and Warburg.
Owen, W. (2002) The horizon. In Fawcett-Tang, R. (ed.), *Mapping: An Illustrated Guide to Graphic Navigational Systems*. Lausanne: Rotovision, 10–14.
Pasquini, M., Lau, L. and Pasquini, M.W. (2008) Jack of all trades'? The negotiation of interdisciplinarity within geography. *Geoforum*, 39(2), 552–560.
Pavlik, A. (2010) 'A special kind of reading game': Maps in children's literature. *International Research in Children's Literature*, 3(1), 28–43.
Pearson, M. and Shank, M. (2001) *Theater/Archeaology*. New York: Routledge.
Peplow, D. (2011) "Oh, I've known a lot of Irish people": Reading groups and the negotiation of literary interpretation. *Language and Literature*, 20(4), 295–315.
Peplow, D. (2016) *Talk About Books: A Study of Reading Groups*. London: Bloomsbury.
Peplow, D., Swann, J., Trimarco, P. and Whiteley, S. (2016) *The Discourse of Reading Groups: Integrating Cognitive and Sociocultural Perspectives*. London: Routledge.
Pernecky, T. (2020) An unintroduction to postdisciplinarity. In Pernecky, T. (ed.), *Postdisciplinary Knowledge*. London: Routledge, 1–20.
Pernecky, T., Munar, A.M. and Wheeller, B. (2016) Existential postdisciplinarity: Personal journeys into tourism, art, and freedom. *Tourism Analysis*, 21(4), 389–401.
Perry, S. (2017) *The Essex Serpent*. London: Serpent's Tail.
Peters, K. (2010) Future promises for contemporary social and cultural geographies of the sea. *Geography Compass*, 49, 1260–1272.
Peters, K. and Steinberg, P. (2015) A wet world: Rethinking place, territory, and time. *Society & Space*. http://societyandspace.org/2015/04/27/a-wet-worldrethinking-place-territory-and-time-kimberley-peters-and-philip-steinberg. Accessed August 8, 2018.
Peters, K., Steinberg, P. and Stratford, E. (eds.). (2018) *Territory Beyond Terra*. London: Rowman and Littlefield.
Philo, C. (2005) Editorial. Spacing lives and lively spaces: Partial remarks on Sarah Whatmore's hybrid geographies. *Antipode*, 824–833.
Piatti, B. (2016) Mapping fiction. In Cooper, D., Donaldson, C. and Murrieta-Flores, P. (eds.), *Rethinking Literary Mapping*. London: Routledge, 88–101.

Piatti, B., Bär, H., Reuschel, A.K., Hurni, L. and Cartwright, W. (2008, February 1–2) Mapping literature: Towards a geography of fiction. In *Art and Cartography Symposium*, Vienna, Austria, 177–192. Berlin: Springer.

Piatti, B. and Hurni, L. (2009) Mapping the ontologically unreal – Counterfactual spaces in literature and cartography. *The Cartographic Journal. Art & Cartography Special Issue*, 46(4), 333–342.

Piatti, B. and Hurni, L. (2011) Editorial. Cartographies of fictional worlds – Special issue. *The Cartographic Journal*, 48(4), 218–223.

Pile, S. (1994) Masculinism: The use of dualistic epistomologies and third spaces. *Antipode*, 263, 5–43.

Pinder, D. (2001) Ghostly footsteps: Voices, memories and walks in the city. *Ecumene: A Journal of Cultural geographies*, 8(1), 1–19.

Pinder, D. (2005) Arts of urban exploration. *Cultural Geographies*, 12, 383–411.

Pink, S. (2007) Walking with video. *Visual Studies*, 22(3), 240–252.

Pocock, D. (1989) Sound and the geographer. *Geography*, 74(3), 193–200.

Pratt, G. (2004) Feminist geographies: Spatialising feminist politics. In Cloke, P., Crang, P. and Goodwin, M. (eds.), *Envisioning Human Geographies*. London: Hodder Arnold, 283–304.

Preston, C. (2003) *Grounding Knowledge: Environmental Philosophy, Epistemology, and Place*. Athens and London: University of Georgia Press.

Pritchard, A. and Morgan, N. (2001) Culture, identity and tourism representation: Marketing Cymru or Wales? *Tourism Management*, 22, 167–179.

Price, A. (2002) *The Life of Rebecca Jones*. London: MacLehose Press.

Prichard, C. (1961) *One Moonlit Night*. London: Penguin.

Prichard, C. (1961) *Un Nos Ola Leuad*. Llandeilo: Y Lolfa.

Pryce, M. (2001) *Aberystwyth, Mon Amour*. London: Bloomsbury.

Pryce, Malcolm (2003) *Last Tango in Aberystwyth*. London: Bloomsbury.

Pryce, Malcolm (2005) *The Unbearable Lightness of Being in Aberystwyth*. London: Bloomsbury.

Pryce, Malcolm (2007) *Don't Cry for Me Aberystwyth*. London: Bloomsbury.

Pryce, Malcolm (2009) *From Aberystwyth with Love*. London: Bloomsbury.

Pryce, Malcolm (2011) *The Day Aberystwyth Stood Still*. London: Bloomsbury.

Pullman, P. (2017) *Daemon Voices: On Stories and Storytelling*. Oxford: David Fickling Books.

Pullman, P. (n.d.) How i found my voice. *Intelligence Squared Podcast*. Accessed January 2024.

Qvistrom, M. (2007) Landscapes out of order: Studying the inner urban fringe beyond the rural-urban divide. *Geografiska Annaler. Series B, Human Geography*, 89(3), 269–282.

Rees, B. (2016) Pigeon, by Alys Conran. *Wales Arts Review*. www.walesartsreview.org/novel-pigeon-by-alys-conran/. Accessed May 2016.

Rees, R. (1980) Historical links between cartography and art. *Geographical Review*, 70(1), 60–78.

Reeve, P. (2018) *Mortal Engines*. London: Scholastic.

Relph, E. (1976) *Place and Placelessness*. London: Pion.

Reuschel, A.K. and Hurni, L. (2011) Mapping literature: Visualisation of spatial uncertainty in fiction. *The Cartographic Journal Cartographies of Fictional Worlds – Special Issue*, 48(4), 293–308.

Reuschel, A.K., Piatti, B. and Hurni, L. (2009, November 15–21) Mapping literature: The prototype of a literary atlas of Europe. In *24th International Cartographic Conference ICC 2009*, Santiago. Chile Santiago de Chile: Military Geographic Institute.

Reuschel, A.K., Piatti, B. and Hurni, L. (2013) Modelling uncertain geodata for the literary atlas of Europe. In Kriz, K., Cartwright, W. and Hurni, L. (eds.), *Understanding Different Geographies, Lecture Notes in Geoinformation and Cartography*. Berlin Heidelberg: Springer-Verlag, 135–157.

Revill, G. (2016) How is space made in sound? Spatial mediation, critical phenomenology and the political agency of sound. *Progress in Human Geography*, 402, 240–256.
Reynolds, A. (2001) *Chasm City*. London: Gollancz.
Rhodes, D. (2007) *Gold*. Edinburgh: Canongate.
Richards, G. (2011) Creativity and tourism: The state of the art. *Annals of Tourism Research*, 38(4), 1225–1253.
Richards, G. and Wilson, J. (2006) Developing creativity in tourist experiences: A solution to the serial reproduction of culture? *Tourism Management*, 27, 1408–1413.
Roberts, K. (1962) *Twyll Heno*. Denbigh: Gwasg Gee.
Roberts, L. (ed.). (2012) *Mapping Cultures*. London: Palgrave Macmillan.
Roberts, L. (2016a) Editorial deep mapping and spatial anthropology. *Humanities*, 5, 5.
Roberts, L. (ed.). (2016b) *Deep Mapping*. Basel: MDPI.
Robertson, S. and Mullen, L. (2022) Navigating through narrative. In Bodenhamer, D., Corrigan, J. and Harris, T. (eds.), *Making Deep Maps: Foundations, Approaches, and Methods*. London: Routledge, 131–147.
Robinson, A.H. (1952) *The Look of Maps: An Examination of Cartographic Design*. Madison: University of Wisconsin Press.
Robinson, M. (2002) Between and beyond the pages: Literature-tourism relationships. In Robinson, M. and Andersen, H.C. (eds.), *Literature and Tourism*. London: Thomson, 39–80.
Rojek, C. (1993) *Ways of Escape: Modern Transformations in Leisure and Travel*. London: Macmillan.
Rose, M. (2002) Landscape and labyrinths. *Geoforum*, 33, 455–467.
Rosenbaum, S. (2001) The map: Shifts, overwrites, overlays. *Performance Research*, 6(2), 54–59.
Rouse, J. (1996) *Engaging Science: How to Understand its Practices Philosophically*. Ithaca, NY: Cornell University Press.
Ryan, M.L. (2013) Transmedial storytelling and transfictionality. *Poetics Today*, 34(3), 361–388.
Ryan, M.L. (2015) Texts, worlds, stories: Narrative worlds as cognitive and ontological concept. In *Narrative Theory, Literature, and New Media: Narrative Minds and Virtual Worlds*, London: Routledge, 11–29.
Ryan, M.L. (2017) Ontological rules. In *The Routledge Companion to Imaginary Worlds*, London: Routledge, 74–81.
Ryan, M.L., Foote, K. and Azaryahu, M. (2016) *Narrating Space/Spatializing Narrative: Where Narrative Theory and Geography*. Columbus: Ohio State University Press.
Saunders, A. (2023) A manual of hinges and a musing on Gruffalo. *Literary Geographies*, 9(2), 253–256.
Sayer, A. (2003) Long live postdisciplinary studies! Sociology and the curse of disciplinary parochialism/imperialism. www.comp.lancs.ac.uk/sociology/papers/Sayer-Long-Live-PostdisciplinaryStudies.pdf
Scarles, C. (2010) Where words fail, visuals ignite. *Annals of Tourism Research*, 37(4), 905–926.
Sharp, J. (2000) Towards a critical analysis of fictive geographies. *Area*, 32(3), 327–334.
Sheller, M. and Urry, J. (2006) The new mobilities paradigm. *Environment and Planning A*, 38, 207–226.
Simmel, G. (1950) *The Sociology of Georg Simmel*. London: Free Press.
Sin, C.H. (2003) Interviewing in "place": The socio-spatial construction of interview data. *Area*, 35(3), 305–312.
Sinclair, I. (2003) *London Orbital*. London: Penguin.
Sinclair, I. (2004) *Downriver*. London: Penguin.
Smith, A. (2024) What is visual poetry. *Language Humanities*. www.languagehumanities.org/what-is-visual-poetry.htm#:~:text=Visual%20poetry%20is%20literary%20verse,reading%20the%20poem%20out%20loud. Accessed January 2024.

Soitu, P. (2021) Turf wars: The territorial claims of spatial literary studies and of literary geography. *Annals of Ovidius University Constanta – Philology*, 2, 115–127.

Soja, E. (1996) *Thirdspace*. Malden: Blackwell.

Soja, E. (1999) Thirdspace: Expanding the scope of the geographical imagination. In Massey, D., Allen, J. and Sarre, P. (eds.), *Human Geography Today*. Cambridge: Polity Press, 260–278.

Solnit, R. (2010) *Infinite City: A San Francisco Atlas*. Berkeley: University of California Press.

Solnit, R. (2019) *Whose Story Is This? Old Conflicts, New Chapters*. London: Granta.

Squire, S.J. (1994) The cultural values of literary tourism. *Annals of Tourism Research*, 21, 103–120.

Squire, S.J. (1996) Literary tourism and sustainable tourism: Promoting 'Anne of Green Gables' in Prince Edward Island. *Journal of Sustainable Tourism*, 4(3), 119–134.

Steinberg, P. (2014) Forword: On Thalassography. In Anderson, J. and Peters, K. (eds.), *Water Worlds: Human Geographies of the Ocean*. Farnham: Ashgate, xiii–xvii.

Steinberg, P. and Peters, K. (2015) Wet ontologies, fluid spaces: Giving depth to volume through oceanic thinking. *Environment and Planning D: Society and Space*, 33(2), 247–264.

Storr, W. (2019) *The Science of Storytelling*. London: William Collins.

Swift, E. (2021) Narratives of the future: Immersive stories/indigenous roots. *Body, Space & Technology*, 21(1), 1–16.

Szerszynski, B. and Galarraga, A. (2013) Geoengineering knowledge: Interdisciplinarity and the shaping of climate engineering research. *Environment and Planning A*, 42(112), 2817–2824.

Tally, R. (2013) *Spatiality*. London & New York: Routledge.

Tally, R. (2017) The reassertion of space in literary studies. In Tally, R. (ed.), *The Routledge Handbook of Literature and Space*. London and New York: Routledge, 1–6.

Tally, R. (2019) Spatial literary studies versus literary geography? *English Language and Literature*, 65(3), 391–408.

Tally, R. (ed.). (2020) *Spatial Literary Studies: Interdiscipinary Approaches to Space, Geography, and the Imagination*. London: Routledge.

Tekgül, D. (2019) Book club meetings as micro public spheres: Translated literature and cosmopolitanism. *Language and Intercultural Communication*, 19(5), 380–392.

Thacker, A. (2017) Critical literary geography. In Tally, R. (ed.), *The Routledge Handbook of Literature and Space*. London and New York: Routledge, 28–38.

Thobo-Carlsen, M. (2016) Walking the museum – Performing the museum. *The Senses and Society*, 11(2), 136–157.

Thomas, R. S. (1968) *Not that he brought flowers*. London: Hart-Davis.

Thumala Olave, M.A. (2021) Exploring the sacrality of reading as a social practice. *American Journal of Cultural Sociology*, 9, 99–114.

Thurgill, J. (2021) Literary geography and the spatial hinge. *Literary Geographies*, 7(2), 152–156.

Thurgill, J. (2023) The spatial hinge: An introduction. *Literary Geographies*, 9(2), 234–236.

Thurgill, J. and Lovell, J. (2019) Expanding worlds: Place and collaboration in (and after) the 'text-as-spatial-event'. *Literary Geographies*, 5(1), 16–20.

Tresize, R. (2000) *In and Out of the Goldfish Bowl*. Cardigan: Parthian.

Tröndle, M., Greenwood, S., Bitterli, K. and van den Berg, K. (2014) The effects of curatorial arrangements. *Museum Management and Curatorship*, 29(2), 140–173.

Tuan, Y.-F. (1975) *Topophilia: A Study of Environmental Perception, Attitudes, and Values*. Englewood Cliffs: Prentice Hall.

Turchi, P. (2004) *Maps of the Imagination: The Writer as Cartographer*. San Antonio: Trinity University Press.

Turkel, W. (2006) Every place is an archive: Environmental history and the interpretation of physical evidence. *Rethinking History*, 10(2), 259–276.

Urquhart, L.L. (2019) Cartographic imaginaries: Interpreting the *Literary Atlas* of Wales. *Swansea Museum*. www.buzzmag.co.uk/cartographic-imaginaries-art-preview/. Accessed July 2019.

Wales Arts Review. (2021) The greatest Welsh novel: A resource. *Wales Arts Review*. www.walesartsreview.org/the-greatest-welsh-novel-a-resource/. Accessed March 2021.

Walford Davies. (n.d.) (Dis)trusting maps. *British Library Transforming Topography*. www.bl.uk/-places/articles/distrusting-maps. Accessed January 2013.

Walford Davies, D. (2012) *Cartographies of Culture: New Geographies of Welsh Writing in English*. Cardiff: University of Wales Press.

Wang, Y. and Ye, Y. (2020) A multimodal analysis of E.E. Cummings' visual poem *L*. *Advances in Vocational and Technical Education*, 2, 6–11.

Warf, B. and Arias, S. (eds.). (2008) *The Spatial Turn: Interdisciplinary Perspectives*. London: Routledge.

Waters, S. (2009) *The Little Stranger*. London: Riverhead Books.

Watson, J.W. (1983) The soul of geography. *Journal Transactions of the Institute of British Geographers*, 8, 385–399.

Watson, N.J. (2006) *The Literary Tourist*. Basingstoke: Palgrave Macmillan.

Watson, V. (2000) *Reading Series Fiction: From Arthur Ransome to Gene Kemp*. London: Routledge Falmer.

Weightman, B.A. (1987) Third world tour landscapes. *Annals of Tourism Research*, 14, 227–239.

Weir, D. (2002) Nevil shute and the landscape of England: An opportunity for literary tourism. In Robinson, M. and Andersen, H. (eds.), *Literature and Tourism: Essays in the Reading and Writing of Tourism*. London: Continuum, 119–142.

Whatmore, S. (2002) *Hybrid Geographies*. London: Sage.

Whiteley, S. and Peplow, D. (2021) Interpreting real and fictional worlds in interaction: A socio-cognitive approach to reading group talk. *Text & Talk*, 41(1), 119–139.

Williams, R. (1960/2006 edition) *Border Country*. Parthian: Cardigan.

Wood, D. (2006) Map art. *Cartographic Perspectives*, 53, 5–14.

Wood, D. (2015) Mapping deeply. *Humanities*, 4, 304–318.

Woodyer, T. and Geoghegan, H. (2012) (Re)enchanting geography? The nature of being critical and the character of critique in human geography. *Progress in Human Geography*, 37(2), 195–214.

Wright, J. (1942) Map makers are human: Comments on the subjective in maps. *Geographical Review*, 32(4), 527–544.

Wynne Jones, D. (1986) *Howl's Moving Castle*. London: Harper Collins.

Yeh, J. and O'Reilly, S. (eds.). (2022) *Creative Writing: A Workbook with Readings*. London: Routledge.

Yiannakis, J. and Davies, A. (2012) Diversifying rural economies through literary tourism: A review of literary tourism in Western Australia. *Journal of Heritage Tourism*, 7(1), 33–44.

Ylofa. (n.d.) Atyniad. www.ylolfa.com/products/9780862439330/atyniad. Accessed January 2024.

Zizek, S. (2004) From purification to subtraction: Badiou and the real. In Hallward, P. (ed.), *Think Again. Alain Badiou and the Future of Philosophy*. London and New York: Continuum, 165–181.

Index

FSC
www.fsc.org
MIX
Papier | Fördert gute Waldnutzung
FSC® C083411

Kolibri 360 GmbH
Ferdinand-Jühlke-Straße 7
99095 Erfurt, Deutschland
produktsicherheit@kolibri360.de